ZIML Competition Book

Varsity Division 2016-2017

Areteem Institute

TITLES PUBLISHED BY ARETEEM PRESS

Cracking the High School Math Competitions (and Solutions Manual) - Covering AMC 10 & 12, ARML, and ZIML
Mathematical Wisdom in Everyday Life (and Solutions Manual) - From Common Core to Math Competitions
Geometry Problem Solving for Middle School (and Solutions Manual) - From Common Core to Math Competitions
Fun Math Problem Solving For Elementary School (and Solutions Manual)
ZIML Math Competition Book Division M 2016-2017
ZIML Math Competition Book Jr Varsity 2016-2017
ZIML Math Competition Book Division E 2016-2017
ZIML Math Competition Book Varsity Division 2016-2017

COMING SOON

ZIML Math Competition Book Division H 2016-2017
Counting & Probability for Middle School (and Solutions Manual) - From Common Core to Math Competitions
Number Theory Problem Solving for Middle School (and Solutions Manual) - From Common Core to Math Competitions

The books are available in paperback and Kindle eBook formats. To order the books, visit

`https://areteem.org/bookstore.`

ZIML Math Competition Book Varsity Division 2016-2017

Edited by   John Lensmire
David Reynoso
Kevin Wang
Kelly Ren

Cover and chapter title photographs by Kelly Ren and Kevin Wang

Copyright © 2018 ARETEEM INSTITUTE
WWW.ARETEEM.ORG

PUBLISHED BY ARETEEM PRESS

ISBN: 1-944863-14-1
ISBN-13: 978-1-944863-14-2

First printing, April 2018.

Contents

Introduction

Each month during the school year, Areteem Institute hosts the online Zoom International Math League (ZIML) competitions. Students can compete in one of five divisions based on their age and mathematical level (details shown on Page 7).

This book contains the problems, answers, and full solutions from the nine ZIML Varsity Competitions held during the 2016-2017 School Year. It is divided into three parts:

1. The complete Varsity ZIML Competitions (20 questions per competition) from October 2016 to June 2017.
2. The solutions for each of the competitions, including detailed work and helpful tricks.
3. An appendix including the topics and knowledge points covered for Varsity, a glossary including common mathematical terms, and answer keys for each of the competitions so students can easily check their work.

The questions found on the ZIML competitions are meant to test your problem solving skills and train you to apply the knowledge you know to many different applications. We hope you enjoy the problems!

About Zoom International Math League

The Zoom International Math League (ZIML) has a simple goal: provide a platform for students to build and share their passion for math and other STEM fields with students from around the globe. Started in 2008 as the Southern California Mathematical Olympiad, ZIML has a rich history of past participants who have advanced to top tier colleges and prestigious math competitions, including American Math Competitions, MATHCOUNTS, and the International Math Olympaid.

The ZIML Core Online Programs, most available with a free account at ziml.areteem.org, include:

- **Daily Magic Spells:** Provides a problem a day (Monday through Friday) for students to practice, with full solutions available the next day.
- **Weekly Brain Potions:** Provides one problem per week posted in the online discussion forum at ziml.areteem.org. Usually the problem does not have a simple answer, and students can join the discussion to share their thoughts regarding the scenarios described in the problem, explore the math concepts behind the problem, give solutions, and also ask further questions.
- **Monthly Contests:** The ZIML Monthly Contests are held the first weekend of each month during the school year (October through June). Students can compete in one of 5 divisions to test their knowledge and determine their strengths and weaknesses, with winners announced after the competition.
- **Math Competition Practice:** The Practice page contains sample ZIML contests and an archive of AMC-series tests for online practice. The practices simulate the real contest environment with time-limits of the contests automatically controlled by the server.
- **Online Discussion Forum:** The Online Discussion Forum

is open for any comments and questions. Other discussions, such as hard Daily Magic Spells or the Weekly Brain Potions are also posted here.

These programs encourage students to participate consistently, so they can track their progress and improvement each year.

In addition to the online programs, ZIML also hosts onsite Local Tournaments and Workshops in various locations in the United States. Each summer, there are onsite ZIML Competitions at held at Areteem Summer Programs, including the National ZIML Convention, which is a two day convention with one day of workshops and one day of competition.

ZIML Monthly Contests are organized into five divisions ranging from upper elementary school to advanced material based on high school math.

- **Varsity:** This is the top division. It covers material on the level of the last 10 questions on the AMC 12 and AIME level. This division is open to all age levels.
- **Junior Varsity:** This is the second highest competition division. It covers material at the AMC 10/12 level and State and National MathCounts level. This division is open to all age levels.
- **Division H:** This division focuses on material from a standard high school curriculum. It covers topics up to and including pre-calculus. This division will serve as excellent practice for students preparing for the math portions of the SAT or ACT. This division is open to all age levels.
- **Division M:** This division focuses on problem solving using math concepts from a standard middle school math curriculum. It covers material at the level of AMC 8 and School or Chapter MathCounts. This division is open to all students who have not started grade 9.

- **Division E:** This division focuses on advanced problem solving with mathematical concepts from upper elementary school. It covers material at a level comparable to MOEMS Division E. This division is open to all students who have not started grade 6.

This problem book features the Varsity Contests. For a detailed list of topics covered for Varsity see p.155 in the Appendix.

About Areteem Institute

Areteem Institute is an educational institution that develops and provides in-depth and advanced math and science programs for K-12 (Elementary School, Middle School, and High School) students and teachers. Areteem programs are accredited supplementary programs by the Western Association of Schools and Colleges (WASC). Students may attend the Areteem Institute through these options:

- Live and real-time face-to-face online classes with audio, video, interactive online whiteboard, and text chatting capabilities;
- Self-paced classes by watching the recordings of the live classes;
- Short video courses for trending math, science, technology, engineering, English, and social studies topics;
- Summer Intensive Camps on prestigious university campuses and Winter Boot Camps;
- Practice with selected daily problems for free, and monthly ZIML competitions at ziml.areteem.org.

The Areteem courses are designed and developed by educational experts and industry professionals to bring real world applications into STEM education. The programs are ideal for students who wish to build their mathematical strength in order to excel academically and eventually win in Math Competitions (AMC, AIME, USAMO, IMO, ARML, MathCounts, Math Olympiad, ZIML, and other math leagues and tournaments, etc.), Science Fairs (County Science Fairs, State Science Fairs, national programs like Intel Science and Engineering Fair, etc.) and Science Olympiad, or purely want to enrich their academic lives by taking more challenges and developing outstanding analytical, logical thinking and creative problem solving skills.

Since 2004 Areteem Institute has been teaching with methodology that is highly promoted by the new Common Core State Standards: stressing the conceptual level understanding of the math concepts, problem solving techniques, and solving problems with real world applications. With the guidance from experienced and passionate professors, students are motivated to explore concepts deeper by identifying an interesting problem, researching it, analyzing it, and using a critical thinking approach to come up with multiple solutions.

Thousands of math students who have been trained at Areteem achieved top honors and earned top awards in major national and international math competitions, including Gold Medalists in the International Math Olympiad (IMO), top winners and qualifiers at the USA Math Olympiad (USAMO/JMO), and AIME, top winners at the Zoom International Math League (ZIML), and top winners at the MathCounts National. Many Areteem Alumni have graduated from high school and gone on to enter their dream colleges such as MIT, Cal Tech, Harvard, Stanford, Yale, Princeton, U Penn, Harvey Mudd College, UC Berkeley, UCLA, etc. Those who have graduated from colleges are now playing important roles in their fields of endeavor.

Further information about Areteem Institute, as well as updates and errata of this book, can be found online at http://www.areteem.org.

Acknowledgments

This book contains the Online ZIML Varsity Problems from the 2016-17 school year. These problems were created and compiled by the staff of Areteem Institute. These problems were inspired by questions from the Areteem Math Challenge Courses, past questions on the ACT/SAT/GRE, past math competitions, math textbooks, and countless other resources and people encountered by the Areteem Curriculum Department in their life devoted to math. We thank all these sources for growing and nurturing our passion for math.

The Areteem staff, including John Lensmire, David Reynoso, Kevin Wang, and Kelly Ren, are the main contributors who compiled, edited, and reviewed this book. Photographs included on the cover and chapter introduction pages are credit to Kelly Ren and Kevin Wang.

Lastly, thanks to all the students who have participated and continue to participate in the Zoom International Math League. Your dedication to the Daily Magic Spells and Monthly Contests makes all of this possible, and we hope you continue to enjoy ZIML for years to come!

1. ZIML Contests

This part of the book contains the Varsity ZIML Contests from the 2016-17 School Year. There were nine monthly competitions, held on the dates found below:

- October 7-8
- November 4-6
- December 2-4
- January 6-8
- February 3-5
- March 3-5
- April 7-9
- May 5-7
- June 2-4

1.1 ZIML October 2016 Varsity

Below are the 20 Problems from the Varsity ZIML Competition held in October 2016.
The answer key is available on p.169 in the Appendix.
Full solutions to these questions are available starting on p.70.

Problem 1
Given ten cards with numbers $1, 2, \ldots, 10$ on them, each with one number. Randomly select 3 of the ten cards. The probability that the largest number is 5 is $K \div \binom{10}{3}$. What is K?

Problem 2
Let x, y be positive integers and satisfy

$$\begin{aligned} xy + x + y &= 69, \\ x^2 y + xy^2 &= 884. \end{aligned}$$

Find the value of $x^2 + y^2$.

Problem 3
Arrange the numbers $1, 2, 3, \ldots, 999$ on a circle, in that order. Start from 1, do the following: skip 1, cross out 2 and 3; skip 4, cross out 5 and 6. Each step skip one number and cross out the next two. Which number is the last one remaining?

Problem 4

Evaluate

$$\left\lfloor \frac{200 \cdot 1}{71} \right\rfloor + \left\lfloor \frac{200 \cdot 2}{71} \right\rfloor + \left\lfloor \frac{200 \cdot 3}{71} \right\rfloor + \cdots + \left\lfloor \frac{200 \cdot 70}{71} \right\rfloor.$$

Problem 5

Attach a positive integer N to the right of any positive integer (for example, attaching 8 to the right of 57, we get 578), if the new number is always divisible by N no matter what the other positive integer is, then call N a "magic number". How many magic numbers are less than 1000?

Problem 6

Consider the statement: the equation $x^2 - mx + 4 = 0$ has at least one root for x in $[-1, 1]$. This statement is true for all m satisfying $|m| \geq L$ for a number L. What is L?

Problem 7

Given 12 distinct beads, how many possible bracelets of 5 beads can be made? Note: rotating or flipping a bracelet does not change the bracelet.

Problem 8

A room has 3 windows of the same shape and size, only one of which is open. Two birds are stuck in the room and can only fly out through the open window. They each fly around in the room and try to find the open window. Assume the first bird does not remember which windows it has tried, and has the same probability to try any of the windows each time it attempts to escape. The second bird however, remembers which window it has already tried, thus it tries each window at most once. The probability that the second bird escapes in fewer tries than the first is $\dfrac{M}{N}$ with $\gcd(M\ N) = 1$. What is M?

Problem 9

Find prime number p such that $5p + 1$ is a perfect cube.

Problem 10

Given that $\odot O_2$ and $\odot O_3$ are externally tangent, and both of them are inside $\odot O_1$ and tangent to $\odot O_1$. Also $O_1 O_2 = 3$, $O_1 O_3 = 6$, and $O_2 O_3 = 7$. Find the radius of $\odot O_2$.

Problem 11

Given that a, b, c, x, y, z are positive reals and

$$a^2 + b^2 + c^2 = 25$$

$$x^2 + y^2 + z^2 = 36$$

$$ax + by + cz = 30$$

The value of $\dfrac{a+b+c}{x+y+z}$ can be written as $\dfrac{R}{S}$ with $\gcd(R, S) = 1$. What is $R + S$?

Problem 12

A trapezoid has area 32, and the sum of its two bases and altitude is 16. Also assume that one of the diagonals is perpendicular to the bases. The length of the other diagonal is $\sqrt{L}$ for an integer L. Find L.

Problem 13

Add parentheses to the expression $1 : 2 : 3$ can have two results: $(1 : 2) : 3 = 1/6$, and $1 : (2 : 3) = 3/2$. If we add parentheses to $1 : 2 : 3 : 4 : 5 : 6 : 7 : 8$, how many different results can we get?

Problem 14

How many solutions does the following equation have?

$$\frac{1}{5} \log_2 x = \sin 5\pi x.$$

Problem 15
Find the remainder when 2^{345} is divided by 400.

Problem 16
The three sides of a triangle are three consecutive integers. The bisector of the largest angle splits the opposite side into two parts, where the shorter part has length $\dfrac{65}{9}$. Find the length of the largest side.

Problem 17
In $\triangle ABC$, $\angle C = 120°$, $\angle B = 30°$. Let D be the point on $\overline{BC}$ such that $\angle ADC = 45°$. Given that $DC = 8$, the length BD is $A\sqrt{B}$ in simplest radical form. What is $A + B$?

Problem 18
The maximum value of $\sin x| + |\cos x|$ can be written as $\sqrt{K}$ for some integer K. What is K?

Problem 19

David is a science fiction fan and he has a collection of 8 sci-fi books written by Isaac Asimov and 7 books by Arthur Clarke. He arranges these books on a shelf in the pattern

$$(A)(C)(A)(C)(A)(C)(A)(C)$$

where each (A) is a group of consecutive Asimov books and each (C) is a group of consecutive Clarke books. Suppose David does not distinguish the books from the same author. In how many ways can he arrange the books according to the pattern above?

Problem 20

In $\triangle ABC$, $\overline{AD}$ is the altitude on side $\overline{BC}$. Given that $\angle A = 45°$, $BD = 2, DC = 3$, find the area of $\triangle ABC$.

1.2   ZIML November 2016 Varsity

Below are the 20 Problems from the Varsity ZIML Competition held in November 2016.
The answer key is available on p.170 in the Appendix.
Full solutions to these questions are available starting on p.80.

Problem 1
Solve $\sqrt{x-4} = x-6$ for real values of x.

Problem 2
Square $ABCD$ has side length 6. Let M be the midpoint of $\overline{CD}$, and O be the circumcenter of $\triangle MAB$. Find the diameter of the incircle of $\triangle OAB$.

Problem 3
Suppose you flip a coin 10 times. The probability of getting more heads than tails can be expressed in the form $\dfrac{p}{q}$ in lowest terms. Find $p+q$.

Problem 4
Let N be the least common multiple of 1, 2, 3, $\ldots$, 1998, 1999, 2000, and 2^k be the maximum power of 2 that divides N. What is k?

Problem 5

The solution of the equation $\log_{\sqrt{3}} x + \log_3 x + \log_9 x = -\dfrac{21}{2}$ can be expressed as a fraction $\dfrac{p}{q}$ in lowest terms. Find $p + q$.

Problem 6

Four identical spherical balls, each of radius 10cm, are glued to the ground so that their centers form the vertices of a square with side length 20cm. Place a fifth ball (with a different radius) on top of the four balls on the ground so the fifth ball is externally tangent to those four balls. Suppose the bottom of the fifth ball is 10cm off the ground. Find the radius of the fifth ball (in cm). (Don't include the unit in your answer)

Problem 7

There are 5 distinct red balls and 4 distinct white balls in a bag. Ed takes a ball from the bag and then places it back. He does this three times. The probability that he gets a red ball twice and a white ball once can be expressed as a fraction $\dfrac{p}{q}$ in lowest terms. Find $p + q$.

Problem 8

How many ordered triples (a, b, c) are there so that $a \cdot b \cdot c = 40500$, where a, b, c are positive integers?

Problem 9

Find the remainder when 2^{987} is divided by 100.

Problem 10

If p, q are positive integers, and the equation (in x)

$$\frac{1}{2}px^2 - \frac{1}{2}qx + 2017 = 0$$

has two prime roots. Find the value of q.

Problem 11

A line passes through point $A(1,0)$ and has exactly one common point with parabola $y = \frac{x^2}{2}$. How many such lines are there?

Problem 12

Let n be a single digit (at most 9). Use the digits $1, 2, \ldots, n$ to form n-digit numbers with no repeating digits, where 2 cannot be adjacent to 1 or 3. Assume there are 2400 such numbers in total. Find the value of n.

Problem 13

Let a, b, c, d (not necessarily real numbers) be the roots of

$$x^4 - 5x^2 + 3x + 1 = 0.$$

Find the value of $a^4 + b^4 + c^4 + d^4$.

Problem 14

Determine the number of ordered pairs of positive integers (m, n) such that the least common multiple of m and n is 210.

Problem 15
Find the sum of all even positive divisors of 10000.

Problem 16
If $\dfrac{1+\tan\alpha}{1-\tan\alpha} = 2017$, then evaluate $\dfrac{1}{\cos 2\alpha} + \tan 2\alpha$.

Problem 17
Assume $x \neq 0$ and

$$\frac{x-2}{x+2} + \frac{x-8}{x+8} = \frac{x-4}{x+4} + \frac{x-6}{x+6},$$

find x.

Problem 18
Given a set of numbers $\{1,2,\ldots,14\}$. Select three numbers a,b,c from this set, such that $b-a \geq 3$ and $c-b \geq 3$. Find the number of ways to select a,b,c.

Problem 19

In triangle ABC, $AB = 15$, $AC = 9$, and $BC = 12$. Circles $\odot O_1$, $\odot O_2$, and $\odot O_3$ are inside $\triangle ABC$, and all have radius r, as shown, and the adjacent circles are externally tangent. Also, $\odot O_1$ is tangent to $\overline{AC}$ and $\overline{BC}$, $\odot O_2$ is tangent to $\overline{BC}$, and $\odot O_3$ is tangent to $\overline{AB}$ and $\overline{BC}$.

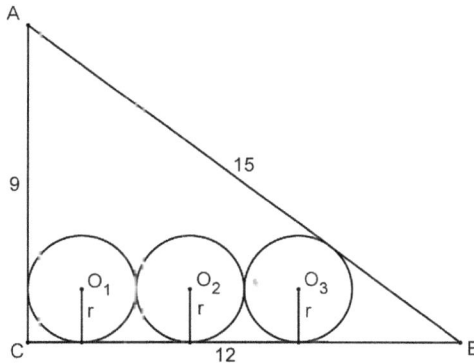

Find the value of r (express your answer as a decimal value, rounded to the nearest tenth).

Problem 20

Let $\lfloor x \rfloor$ represent the largest integer not exceeding x. In the sequence

$$\left\lfloor \frac{1^2}{2000} \right\rfloor, \left\lfloor \frac{2^2}{2000} \right\rfloor, \ldots, \left\lfloor \frac{2000^2}{2000} \right\rfloor,$$

how many distinct values are there?

1.3 ZIML December 2016 Varsity

Below are the 20 Problems from the Varsity ZIML Competition held in December 2016.
The answer key is available on p.171 in the Appendix.
Full solutions to these questions are available starting on p.90.

Problem 1
What is the sum of all positive integers n for which $(n+1)$ divides $(n^2 + 1)$?

Problem 2
Five couples get together for dinner. They sit around a circular table (with 10 seats) with each couple sitting next to their partner. If we only care about the seating arrangement around the table (not the actual seats people sit in—so if one seating arrangement can be rotated to become identical to another arrangement, the two arrangements are considered the same), how many distinct arrangments are there?

Problem 3
Points A, B, C, D lie in the plane. Suppose C lies between A, B on $\overline{AB}$ while D is not on line $\overleftrightarrow{AB}$. If $CD = BC = 3, AC = 6$, and $BD = 4$, what is AD?

Problem 4
Let $x, y, z > 1$ be such that $\log_{xy} z = 2$ and $\log_z(x/y) = 1$. What is $\log_y z$?

Problem 5

A five-digit number N consists of 5 distinct nonzero digits, and N equals the sum of all possible 3-digit numbers made up of 3 of its 5 digits. Find the sum of all such 5-digit numbers N.

Problem 6

A spherical ball was floating on the lake when the lake was frozen. The ball was removed without breaking the ice. A hole was left on the surface of the ice, with diameter 24 and depth 8. Find the radius of the ball.

Problem 7

Find the minimum value of

$$\frac{a}{b+c} + \frac{b}{c+a} + \frac{c}{a+b}$$

if $a, b, c > 0$. Give your answer as a decimal rounded to the nearest tenth.

Problem 8

Find the remainder when $10^{10} + 10^{10^2} + 10^{10^3} + \cdots + 10^{10^{10}}$ is divided by 7.

Problem 9

You are playing darts. The probability you hit the target is 40% on any given throw. Let X be the number of throws it takes until you hit the target. The probability that X is even is $K\%$ for a number K. What is K?

Problem 10
Let $z \neq 1$ be a complex number satisfying $z^7 = 1$. Find the value of

$$\frac{z}{1-z^2} + \frac{z^2}{1+z^4} + \frac{z^3}{1+z^6}.$$

Problem 11
Given $\triangle ABC$ with $\cos B = \dfrac{24}{25}$, $\cos C = \dfrac{12}{13}$, $\sin A = \dfrac{P}{Q}$ where $\gcd(P,Q) = 1$. What is $Q - P$?

Problem 12
After school, Jack and Jill plan to meet up for dinner. They both randomly arrive between 5:00 and 6:00 PM. Jill will wait up to 15 minutes for Jack. Jack will usually wait up to 30 minutes for Jill, but if he is running late and arrives after 5:45 he will leave without waiting for Jill. The probability that Jack and Jill have dinner together is $\dfrac{P}{Q}$ where $\gcd(P,Q) = 1$. What is $P + Q$?

Problem 13
Consider the prime factorization of the sum:

$$\binom{2016}{0} - \binom{2016}{2} + \binom{2016}{4} - \binom{2016}{6} + \cdots$$
$$\cdots + \binom{2016}{2012} - \binom{2016}{2014} + \binom{2016}{2016}.$$

How many powers of 2 appear in this prime factorization? (For example, 3 powers of 2 appear in the prime factorization of $24 = 2^3 \cdot 3$.)

Problem 14
In a right triangle $\triangle ABC$ suppose $\angle B = 90°$ and $\angle C = 30°$. Suppose point D is on $\overline{BC}$ with $\angle ADB = 45°$ and $DC = 10$. Find the length of AB rounded to the nearest integer.

Problem 15
How many distinct real roots does

$$(x^2 + x - 2)^2 + (2x^2 - 5x + 3)^2 = (3x^2 - 4x + 1)^2$$

have?

Problem 16
Suppose a sequence has recursive definition $a_1 = k$ for $k \geq 1$ and $a_{n+1} = 4^{a_n} \pmod{11}$. Find the sum of all possible a_{100}'s.

Problem 17

Let O be the circumcenter of $\triangle ABC$. Through A construct a line tangent to $\odot O$ and intersecting the extension of $\overrightarrow{BC}$ at D. From B and C construct lines perpendicular to $\overline{AD}$ with feet M and N respectively. Assume that $CN = 10, BC = 10, AD = 10\sqrt{6}$. The area of trapezoid $MBCN$ is $p\sqrt{q}$ in simplest radical form. What is $p + q$?

Problem 18

Find the smallest prime factor of $29! - 1$.

Problem 19

a, b, c, d are real numbers such that $a + b + c + d = 0$. Calculate

$$\frac{a^3 + b^3 + c^3 + d^3}{abc + bcd + cda + dab}.$$

Problem 20

How many ways are there to write 201 as the sum of three non-negative integers (we care about the order of the numbers) if all three numbers must be different?

1.4 ZIML January 2017 Varsity

Below are the 20 Problems from the Varsity ZIML Competition held in January 2017.
The answer key is available on p.172 in the Appendix.
Full solutions to these questions are available starting on p.99.

Problem 1
Find the remainder when $68^{50} + 77^{65}$ is divided by 13.

Problem 2
Suppose in triangle ABC we have $\angle A = 45°$, $AB = 3$ and $AC = 2\sqrt{2}$. What is $\tan B$ rounded to the nearest integer?

Problem 3
The equation $(x^2 + x + 1) + (x^2 + 2x + 3) + (x^2 + 3x + 5) + \cdots + (x^2 + 25x + 49) = 6375$ has one positive and one negative solution. What is the negative solution? Round your answer to the nearest tenth if necessary.

Problem 4
Seven identical rooks are placed on an 8×8 chessboard to that they are in different rows and columns. How many such arrangements are there?

Problem 5
A four digit number minus the sum of its digits is $\overline{31d2}$. What is d?

Problem 6

In quadrilateral $ABCD$, $\angle ABC = 135°$, $\angle BCD = 120°$, $AB = \sqrt{6}$, $BC = 5 - \sqrt{3}$, and $CD = 6$. Find AD^2.

Problem 7

Consider solutions to the system of congruences below:

$$x \equiv 4 \pmod{11}, \quad x \equiv 3 \pmod{17}.$$

What is the sum of all solutions x satisfying $|x| < 1000$.

Problem 8

As shown in the diagram, $\overline{AB}$, $\overline{AC}$ are two chords of circle O, and $AB = AC$. Through C construct tangent line to $\odot O$, intersecting the extension of $\overline{BA}$ at D. Extend $\overline{CA}$, and $\overline{DE} \perp \overline{CA}$ at E.

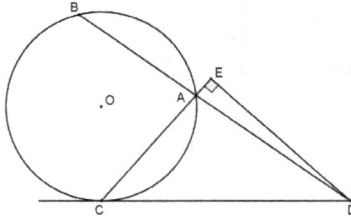

If $CE = 3$, what is BD?

Problem 9

The equation $x^3 - 5x^2 + 7x + 1 = 0$ has three different complex roots a, b, and c. Find $a^3 + b^3 + c^3$.

Problem 10
Randomly put 4 distinct balls into 4 numbered boxes. The probability that there is exactly one empty box can be written as $\dfrac{P}{256}$ for an integer P. What is P?

Problem 11
In trapezoid $ABCD$, $\overline{AB} \parallel \overline{DC}$, with $AB < CD$ and $AD = BC$. Let O be the intersection of $\overline{AC}$ and $\overline{BD}$. Given that $AC = 2$ and $[ABCD] = \sqrt{2}$, find $\angle AOB$. Express your answer in degrees. ($[ABCD]$ means the area of $ABCD$.)

Problem 12
Jane and George play a game. They alternate rolling a fair 6-sided die until one of them gets a 6. The person who rolls the 6 wins. If Jane goes first, the probability she (eventually) wins the game is $\dfrac{P}{Q}$ with $\gcd(P,Q) = 1$. What is $P+Q$?

Problem 13
How many solutions are there to $a \cdot b \cdot c = 40500$, where a, b, c are integers?

Problem 14
The inequality $\dfrac{|3x^2 + kx + 8|}{x^2 - x + 1} \leq 8$ holds for all $x \in \mathbb{R}$. Find the maximum possible value of k.

Problem 15

Find the sum of all the REAL roots of

$$-2x^4 + 2x^3 - 3x^2 - 3x + 9 = 0.$$

Problem 16

Find the constant term in the expansion of

$$\left(\sqrt{x} + \frac{1}{\sqrt{x}} - 2\right)^5.$$

Problem 17

Find the largest integer n such that $n^2 + 2018n$ is a perfect square.

Problem 18

Square $DEFG$ is inscribed in $\triangle ABC$, where $\overline{DE}$ is on $\overline{AB}$, F and G are on $\overline{BC}$ and $\overline{AC}$ respectively. Given that $AB = 40$, altitude $CH = 24$, find the side length of $DEFG$.

Problem 19

Consider only the solutions to

$$x^7 + x^6 + x^5 + x^4 + x^3 + x^2 + x + 1 = 0$$

that are not real. The sum of the squares of these solutions can be written as $A + Bi$ for integers A, B. What is $A + B$?

Problem 20

Find the sum of all primes p such that $p \mid 29^p + 1$.

1.5   ZIML February 2017 Varsity

Below are the 20 Problems from the Varsity ZIML Competition held in February 2017.
The answer key is available on p.173 in the Appendix.
Full solutions to these questions are available starting on p.108.

Problem 1
What is the product of all the real roots of $x^4 + 8x^3 - 8x - 64 = 0$?

Problem 2
3 numbered red balls and 10 identical green balls are placed in a row. If there are at least two green balls separating each of the red balls, how many ways are there to arrange the 13 balls?

Problem 3
Let N be the largest solution less than 5000 to

$$8x \equiv 55 \pmod{2017}.$$

What is N?

Problem 4

Suppose $\overline{AB}$ and $\overline{CD}$ are two chords of equal length, intersecting at E as in the diagram below.

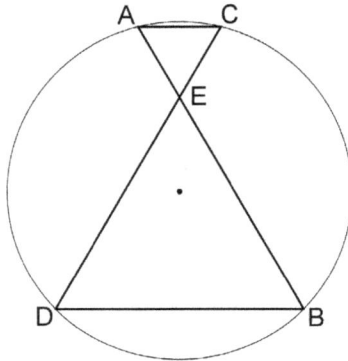

Suppose $\triangle ABE$ and $\triangle CDE$ are equilateral, with $AC = 18$ and $BD = 54$. Then the area of the circle is $K\pi$. What is K, rounded to the nearest integer if necessary?

Problem 5

Assume $(3x-1)^7 = a_7x^7 + a_6x^6 + \cdots + a_1x + a_0$, find the value of $a_0 + a_2 + a_4 + a_6$.

Problem 6

Three mutually-tangent spheres, each with radius 10 inches, rest on the ground. These spheres are fixed in place and a fourth identical sphere is added (on the top) so that the fourth sphere is tangent to all the other spheres. If the total height of the shape formed by the four spheres is H, what is $\lfloor H \rfloor$? Here $\lfloor x \rfloor$ represents the greatest integer not exceeding x.

Problem 7

The inequality $\sqrt{x^2 + 4x} \leq 4 - \sqrt{16 - x^2}$ has finitely many integer solutions. What is the sum of all these solutions?

Problem 8

Suppose that you have a stick that is 10 inches long. Suppose the stick is broken at two random points, which divides the stick into three pieces. The probability that the middle piece of the stick is longer than two inches is $K\%$, where K is an integer. What is K?

Problem 9

Equilateral triangle ABC of side length 2 is drawn. Three squares external to the triangle, $ABDE$, $BCFG$, and $CAHI$, are drawn. The area of the smallest triangle that covers these squares can be written as $R + S\sqrt{T}$ in simplest radical form. What is $R + S + T$?

Problem 10

The repeating decimal $0.357357357\ldots$ can be written as $\frac{x}{y}$, where x and y are relatively prime. What is x?

Problem 11

Old McDonald went to the Market and bought 100 chickens for exactly 100 dollars, among which roosters cost 5 dollars each, hens cost 3 dollars each, and three baby chicks cost 1 dollar. What is the maximum number of roosters that Old McDonald could have bought?

Problem 12

The domain of $\log_3(1 + \log_{1/3}(x - 4))$ is an open interval of length L for an integer L. What is L?

Problem 13

Keith gets off work everyday at 5pm. He can choose to take the bus or subway to get home. From his experience, the probabilities of getting home at certain time range taking either subway or bus is in the following table:

Time to reach home	5:35--5:39	5:40--5:44	5:45--5:49	5:50--5:54	later than 5:54
Subway	0.10	0.25	0.45	0.15	0.05
Bus	0.30	0.35	0.20	0.10	0.05

Today, he decided to flip a coin to choose between bus and subway, and then he arrived home at 5:47. The probability he took the subway is $\dfrac{N}{M}$ with $\gcd(N,M) = 1$. What is $M - N$?

Problem 14

Consider the sum

$$N = \binom{15}{0}^2 + \binom{15}{1}^2 + \binom{15}{2}^2 + \cdots + \binom{15}{15}^2.$$

What is the largest prime that divides N?

Problem 15

Let $x = \dfrac{1}{3 - \sqrt{7}}$. Find $\lfloor x \rfloor + (1 + \sqrt{7})\{x\}$. Here $\lfloor x \rfloor$ represents the greatest integer not exceeding x and $\{x\} = x - \lfloor x \rfloor$.

Problem 16

Let ABC be an isosceles right triangle with $\angle C = 90°$. Let P be a point inside the triangle such that $AP = 3, BP = 5$, and $CP = 2\sqrt{2}$. Then the area of triangle ABC is $\dfrac{N}{M}$ with $\gcd(N, M) = 1$. What is $N + M$?

Problem 17

Consider solutions (x, y) to

$$\begin{cases} x^3 + y^3 - 18xy &= 0 \\ x^2 + y^2 - 20x &= 0 \end{cases}$$

Clearly $(0, 0)$ is a solution. There is one other solution (p, q) with p and q both integers. What is $p + q$?

Problem 18

Calculate the remainder when

$$13^{19^{16}}$$

is divided by 17.

Problem 19

Suppose you have 10 identical balls. How many ways are there to put them in 7 numbered boxes so that at least one of the boxes gets at least 4 balls?

Problem 20

Given that $x, y \in \left[-\dfrac{\pi}{4}, \dfrac{\pi}{4}\right]$, and

$$\begin{cases} x^3 + \sin x - 16 &= 0, \\ 4y^3 + \sin y \cos y + 8 &= 0. \end{cases}$$

Find the value of $(x + 2y)^3 - \cos(x + 2y)$, rounded to the nearest integer if necessary.

1.6 ZIML March 2017 Varsity

Below are the 20 Problems from the Varsity ZIML Competition held in March 2017.
The answer key is available on p.174 in the Appendix.
Full solutions to these questions are available starting on p.118.

Problem 1
Suppose you have a trapezoid $ABCD$ with $\overline{AB}$ parallel to $\overline{CD}$. Let E be the intersection of the diagonals. Suppose $AB = 10, CD = 15$ and $\triangle ADE$ has area 24. Find the area of $ABCD$.

Problem 2
Find the remainder when $31^{999} + 65^{100}$ is divided by 32.

Problem 3
Find the coefficient of x^{11} in $\left(x^5 + x + 2\right)^{10}$.

Problem 4
Let x_1 and x_2 be the two roots of the equation $4x^2 - 8x + k = 0$. Suppose further that $\dfrac{1}{x_1} + \dfrac{1}{x_2} = \dfrac{8}{3}$. Find k.

Problem 5
The equation $|x^2 + 6x - 1| = |(x+3)^2 - 4|$ has solutions $A \pm \sqrt{B}$, where A and B are integers. What is $|A| + |B|$?

Problem 6
What is the smallest integer that has exactly 60 factors?

Problem 7
Suppose in triangle ABC we have $\angle A = 45°$, $AB = 3$ and $AC = 2\sqrt{2}$. What is $\tan(B)$?

Problem 8
Suppose you have 6 numbered red cards and 15 identical black cards. How many ways are there to arrange the cards so that there is at least two black cards between any two red cards?

Problem 9
As shown in the diagram, three congruent squares are drawn next to each other. Angles $\angle 1$ and $\angle 2$ are as labeled.

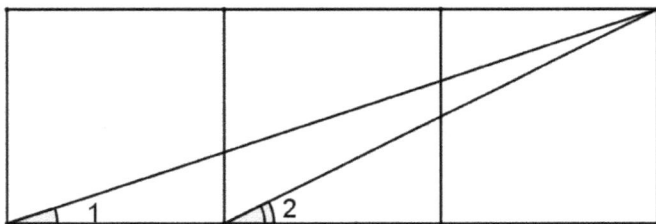

Find $\angle 1 + \angle 2$. Give your answer in degrees.

Problem 10

Adding parentheses to the expression $1/2/3$ can yield two results: $((1/2)/3) = (1/2)/3 = 1/6$, and $1/(2/3) = (1)/(2/3) = 3/2$. If we add parentheses to $1/2/3/4/5/6/7/8$, how many different results can we get?

Problem 11

For how many integer values is $\log(x-20) + \log(30-x) < 1$?

Problem 12

The product of all distinct positive divisors of 60^4 that are perfect squares can be expressed as 60^K for an integer K. What is K?

Problem 13

Suppose you first toss a fair coin. You then roll a fair die once if you get heads and twice if you get tails. Given that the sum of your rolls is 6 (if you only roll once, that result is considered the sum), let the probability your coin came up tails be $\frac{P}{Q}$, where $\gcd(P,Q) = 1$. What is $Q - P$?

Problem 14

ABC is a triangle with integer side lengths. Extend $\overline{AC}$ beyond C to point D such that $CD = 212$. Similarly, extend $\overline{CB}$ beyond B to point E such that $BE = 208$ and $\overline{BA}$ beyond A to point F such that $AF = 204$. If triangles CBD, BAE, and ACF all have the same area, what is the minimum possible area of triangle ABC?

Problem 15

Let $\lfloor x \rfloor$ represent the greatest integer not exceeding x. Consider solutions to $\lfloor -1.77x \rfloor = \lfloor -1.77 \rfloor x$. What is the sum of all such solutions?

Problem 16

Let x be a complex number satisfying $x^2 + x + 1 = 0$, calculate

$$\left(x + \frac{1}{x}\right)^2 + \left(x^2 + \frac{1}{x^2}\right)^2 + \cdots + \left(x^{27} + \frac{1}{x^{27}}\right)^2.$$

Problem 17

How many 5-digit numbers $\overline{abcde}$ are there such that the sum of $\overline{abcd}$ and $\overline{bcde}$ is divisible by 11? Note that 01234 is not a 5-digit number.

Problem 18

Suppose $ABCD$ is a square with side length 2. Suppose a line containing A intersects side CD at E and the line extending BC at F. Find the value of $\frac{1}{AE^2} + \frac{1}{AF^2}$. Express your answer as a decimal.

Problem 19

Suppose you have the numbers $\{0, 1, 2, 3, 4, 5\}$. How many 6-digit numbers can be formed with 1 next to 0 or 2? You many use each number exactly once.

Problem 20

Assume $a, b, c > 0$ and let

$$f(a,b,c) = \frac{a}{b+c} + \frac{4b}{c+a} + \frac{5c}{a+b}.$$

If L is the smallest integer such that $f(a,b,c) = L$ for some a,b,c, what is L?

1.7   ZIML April 2017 Varsity

Below are the 20 Problems from the Varsity ZIML Competition
held in April 2017.
The answer key is available on p.175 in the Appendix.
Full solutions to these questions are available starting on p.127.

Problem 1
Three circles with radii $8, 8, 9$ are externally tangent to each other,
and they are all inside a big circle $\odot O$ and tangent to $\odot O$. Find
the radius of $\odot O$.

Problem 2
Assume $m > 1$. Given that four numbers, 2836, 4582, 5164, and
6522 have the same remainder $r > 0$ when divided by m. Find
the sum of all possible values of r.

Problem 3
21 identical basketball tickets are to be distributed to 4 groups of
students, where any group cannot get more tickets than the other
three groups combined. In how many ways can this be done?

Problem 4
Find all the real solutions to $2\log_{10}(x)\log_{10}(\sqrt{x}) = \log_{10}(100x)$.
What is the product of these solutions?

Problem 5

Let *ABCD* be a parallelogram as in the diagram, with *E* the midpoint of $\overline{BC}$.

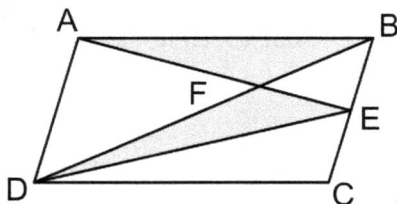

Find the combined area of the shaded regions $\triangle ABF$ and $\triangle DEF$ if the entire parallelogram has area 72.

Problem 6

Consider the 9 numbers

$$1,2,3,11,22,33,111,222,333.$$

How many arrangements of these numbers are there so that the product of any two consecutive numbers is divisible by 3?

Problem 7

Find the greatest common divisor of 20162017 and 20152016.

Problem 8

Let a,b,c be the roots of x^3+4x^2-3x-3. What is $a^3+b^3+c^3$?

Problem 9

Let an equilateral triangle and its incircle be given. An arc on the incircle has the same length as the side length of the equilateral triangle. The angular measure, in radians, of the arc can be expressed as $\dfrac{A\sqrt{B}}{C}$ where $A, B, C > 0$ are integers where B is not divisible by the square of any prime, and $\gcd(A,C) = 1$. What is $A + B + C$?

Problem 10

There are 30 marbles in a bag, labeled 1 through 30. Suppose the weight of marble with label n is equal to $n^2 - 20n + 103$ grams. Select the marbles from the bag with equal probability without considering the weights. If two marbles are selected at once, the probability the two marbles have the same weight can be written as a simplified fraction $\dfrac{P}{Q}$. What is $P + Q$?

Problem 11

Carrie writes the numbers $1, 2, 3, \ldots, 150$ (in order) on the board. Starting from the front, she erases one number, keeps one number, erases one number, etc. until she works through the entire list once. She then starts back at the beginning of the (now smaller) list, erasing one number, keeping one number, etc. This continues until there is only one number remaining on the board. What is this number?

Problem 12

Consider integers a such that the equation (in x)

$$\sqrt{2x-4} - \sqrt{x+a} = 1$$

has exactly one integer root. Find the largest such a less than 10.

Problem 13

Calculate

$$\binom{5}{0} + \binom{6}{1} + \binom{7}{2} + \binom{8}{3} + \cdots + \binom{15}{10}.$$

Problem 14

Find the remainder when $3^3 + 3^{3^2} + 3^{3^3} + \cdots + 3^{3^8}$ is divided by 11.

Problem 15

Let a, b be complex numbers with $|a| = |b| = 2$ and $a + b + 2 = 0$. Consider the imaginary part of $a - b$, $\mathrm{Im}(a - b)$. Then we can write $|\mathrm{Im}(a - b)| = \sqrt{S}$ for an integer S. What is S?

Problem 16

Start with a triangular right prism $ABC - DEF$ (so A is directly 'above' D, B directly 'above' E, etc.). Divide the prism into four parts using the planes through points A, B, F and D, E, C. The ratio of the volume of the largest part to the smallest part is $K : 1$. What is K? If necessary, round your decimal to the nearest tenth.

Problem 17

Let $M = 1 + 2 + 3 + \cdots + 2016$. Find the remainder when 2016! is divided by M.

Problem 18

One solution to the cubic $-x^3 - \sqrt{2}x^2 + \sqrt{2}x + x + \sqrt{2} + 2 = 0$ is of the form $\sqrt{a + \sqrt{b}}$ (with a, b non-zero integers and b not a square). What is $a + b$?

Problem 19

Suppose you have three sticks. One stick has length 1 foot. The second stick has a randomly assigned length from the interval $[0, 2]$ (of real numbers). The third stick has a randomly assigned length from the interval $[0, 3]$ (again of real numbers). What is the probability you can make a triangle using the three sticks? Express your answer as $L\%$, where L is rounded to the nearest integer if necessary.

Problem 20

As shown in the diagram, in isosceles right triangle ABC, $\angle ACB = 90°$. Points D and E are on side $\overline{AB}$, $AD = 1.7$, $BE = 2.64$, and $\angle DCE = 45°$.

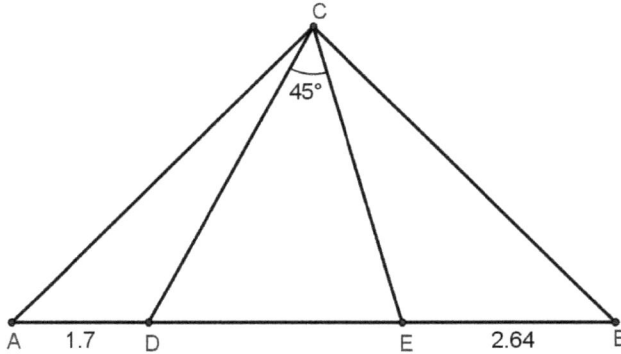

Find the length of $\overline{DE}$.

1.8   ZIML May 2017 Varsity

Below are the 20 Problems from the Varsity ZIML Competition
held in May 2017.
The answer key is available on p.176 in the Appendix.
Full solutions to these questions are available starting on p.136.

Problem 1
Given square $ABCD$ with side length 20, use $\overline{AB}$ as diameter
and construct a semicircle inside the square. From C construct
a tangent line $\overline{CF}$ to the semicircle with the tangent point E,
intersecting $\overline{AD}$ at F. Find DF.

Problem 2
Find the remainder when the sum $n = 2 + 212 + 21212 + \cdots +$
21212121212 (the last term has five 21's and a 2) is divided by
11.

Problem 3
Find the number of 5-digit odd numbers whose digits are all
distinct. Remember that 01357 is NOT a 5-digit number.

Problem 4
If x_1 and x_2 are integer roots of the equation $x^2 + mx + 2 - n = 0$,
and $(x_1^2 + 1)(x_2^2 + 1) = 10$, how many possible pairs (m, n) are
there?

Problem 5

Suppose the altitudes of a triangle are in ratio $10 : 10 : 13$ and the triangle has perimeter 72. What is the area of the triangle? Round your answer to the nearest integer if necessary.

Problem 6

Put 11 distinct balls into 3 distinct boxes, so that each box contains an odd number of balls. How many ways are there to do this?

Problem 7

Find the smallest prime number p such that $5p + 1$ is a perfect cube.

Problem 8

For how many integers n is $\log_{100}(\log_{1/10}(\log_{10}(n^2 - 1)))$ defined?

Problem 9

In $\triangle ABC$, $AB = 71$ and $AC = 84$. Using A as the center and AB as radius, draw a circle to intersect $\overline{BC}$ at D where D is between B and C. Given that the lengths of $\overline{BD}$ and $\overline{DC}$ are both integers, compute BC.

Problem 10

Consider solutions to the system

$$
\begin{aligned}
x &\equiv 2 \quad (\text{mod } 5)\\
x &\equiv 4 \quad (\text{mod } 9)\\
x &\equiv 6 \quad (\text{mod } 11).
\end{aligned}
$$

What is the largest such solution with $x < 10,000$?

Problem 11

If $x^2 + y^2 \le 2$, then maximum value of $|x^2 - 2xy - y^2|$ can be written as $\sqrt{M}$ for an integer M. What is M?

Problem 12

Suppose you roll a fair six-sided die. You then flip a coin the number of times shown on the die. Given that you get exactly 5 heads (no information is given about the number of tails), the probability you rolled a 5 can be expressed as $\dfrac{P}{Q}$ for positive integers P, Q with $\gcd(P, Q) = 1$. What is $P + Q$?

Problem 13

What is the sum of all the integer solutions to

$$(x+1)(x+3)(x+5)(x+7) + 15 = 0?$$

Problem 14

Form a non-regular tetrahedron with an equilateral triangle of side length 1 as the base such that the other 3 edges have length 2. The volume of this tetrahedron can be written in the form $\dfrac{\sqrt{P}}{Q}$ for integers P, Q. What is $P + Q$?

Problem 15

A multiple-choice test had 3 problems, each of which had 4 choices. Only one choice was correct for each problem. David and John made random guesses for each of the problems. At least one of their choices was different. The probability that one of them got all three correct answers can be expressed as $\dfrac{N}{M}$ with N, M positive integers and $\gcd(N, M) = 1$. What is $M - N$?

Problem 16

Consider all ordered triples (x, y, z) of prime numbers satisfying the equation $x(x + y)(y + z) = 140$. How many such triples are there?

Problem 17

Calculate

$$\frac{1}{1024}\left(1 \cdot \binom{10}{1} + 2 \cdot \binom{10}{2} + 3 \cdot \binom{10}{3} + \cdots + 10 \cdot \binom{10}{10}\right).$$

Express your answer as a number rounded to the nearest integer if necessary.

Problem 18

Let $\triangle ABC$ be equilateral triangle, and P be a point in the interior of $\triangle ABC$. Assume that $\angle APB = 113°$, $\angle BPC = 123°$. In fact, AP, BP, CP can be used as the three side lengths of a triangle. For this triangle (with side lengths AP, BP, CP) what is the largest angle (measured in degrees)?

Problem 19

Assume $f(x)$ and $g(x)$ are inverse functions, and $f(x) + f(-x) = 4$. Find the value of $g(x-1) + g(5-x)$.

Problem 20

Find the last digit of $47^{47^{\cdot^{\cdot^{\cdot^{47}}}}}$, where there are 47 total 47s.

1.9 ZIML June 2017 Varsity

Below are the 20 Problems from the Varsity ZIML Competition held in June 2017.
The answer key is available on p.177 in the Appendix.
Full solutions to these questions are available starting on p.145.

Problem 1

Suppose you have a square $ABCD$ with side length 2. Let E be a randomly chosen point inside the square. Then the probability that triangle $\triangle ABE$ is obtuse can be written as $K\%$. What is K, rounded to the nearest integer?

Problem 2

How many of the first 100 Fibonacci numbers are multiples of 3? Recall the Fibonacci numbers start $1, 1, 2, 3, 5, \ldots$.

Problem 3

Find the sum of the 13th powers of the 13 roots of

$$x^{13} - 13x + 13 = 0.$$

Problem 4

Assume $(x-1)^2 \mid [x^4 + (m+n)x^3 + (m-n)x^2 + (m^2 + 2n - 1)x + m + 2]$. There is one such pair of m, n with $m \neq 0$ and $n \neq 0$. For this pair, find $m \times n$.

Problem 5

Four non-overlapping regular plane polygons all have sides of length 1. The polygons meet at a point A in such a way that the sum of the four interior angles at A is $360°$. Among the four polygons, two are squares and one is a triangle. What is the perimeter of the entire shape?

Problem 6

Consider $n!$ expressed in base 8. For example, the base 8 representations of $n!$ for $n = 1, 2, 3, 4$ are $1_8, 2_8, 6_8, 30_8$ (The $_8$ denotes that the number is represented in base 8). For how many n will such a representation end in 16 zeros?

Problem 7

Find the number of positive integers between 1 and 1000000 (inclusive) that are neither perfect squares, nor perfect cubes, nor perfect fifth powers.

Problem 8

Consider numbers n such that the product of all the divisors of n is n^4. (For example the product of all divisors of 10 is $1 \times 2 \times 5 \times 10 = 100 = 10^2 \neq 10^4$ so $n = 10$ does not work.) What is the smallest such $n > 1$?

Problem 9

Let $a \neq b$ be positive real numbers that are solutions to $\log_4(x^2) \cdot \log_2(x) = 1 + 3\log_2(x)$. What is $a \cdot b$?

Problem 10

Bob and Jay play a game with a fair coin and a fair six sided die. Bob flips the coin. If the result of the coin flip heads, then Jay rolls the die twice. If the result of the coin flip is tails, then Jay rolls the die three times. The probability that the sum of the rolls is 7 can be written as $\dfrac{P}{Q}$ with P, Q positive integers and $\gcd(P, Q) = 1$. What is $Q - P$?

Problem 11

Given acute triangle ABC, draw circles using $\overline{AB}$, $\overline{BC}$, $\overline{CA}$ as diameters, as shown in the figure.

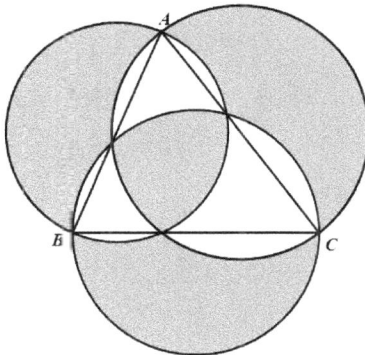

Suppose the total area of the shaded regions outside $\triangle ABC$ is 225, and the area of the shaded region inside $\triangle ABC$ is 45, find the area of the triangle.

Problem 12

Consider n such that $2^n - 1$ is divisible by 3, 11 and 31. What is the sum of all such n with $0 < n < 1000$?

Problem 13

In trapezoid $ABCD$, $\overline{AD} \parallel \overline{BC}$, and let points E, G be on $\overline{AB}$, points F, H be on $\overline{DC}$ such that $AE = EG = GB$, and $DF = FH = HC$. Given that $AD = 20$, and $BC = 29$, find the length of $\overline{EF}$.

Problem 14

4 can be written as a sum of ones and twos in 5 ways:

$$4 = 1+1+1+1 = 1+1+2 = 1+2+1 = 2+1+1 = 2+2.$$

How many ways are there to write 14 as a sum of ones and twos?

Problem 15

Find the remainder when $7^{7^{7^7}}$ is divided by 96.

Problem 16

Suppose $\sin(\theta) = \dfrac{5}{13}$ where $\dfrac{\pi}{2} < \theta < \pi$. Then $\tan(2\theta)$ can be written as $\dfrac{P}{Q}$ for integers P, Q with $Q > 0$ and $\gcd(P, Q) = 1$. What is $P + Q$?

Problem 17

Let ABC be a triangle. Points D and E lie on sides BC and CA, respectively, such that $\angle BAD = \angle DAC$ and $\angle CBE = \angle EBA$. Suppose that line DE bisects $\angle ADC$. Compute $\angle CAB$ in degrees.

Problem 18

Let k be the greatest real root to the equation $x^4 + (x-4)^4 = 626$. What is k? Round your answer to the nearest tenth if necessary.

Problem 19

On one straight line there are 2016 points. Use these points as end points for line segments. At least how many distinct midpoints of these segments are there?

Problem 20

Let x be a complex number satisfying $x^2 + x + 1 = 0$, calculate

$$\left(x^{2016} + \frac{1}{x^{2016}}\right)^3 + \left(x^{2017} + \frac{1}{x^{2017}}\right)^3 + \left(x^{2018} + \frac{1}{x^{2018}}\right)^3.$$

2. ZIML Solutions

This part of the book contains the official solutions to the problems from the nine Varsity ZIML Contests from the 2016-17 School Year.

Students are encouraged to discuss and share their own methods to the problems using the Discussion Forum on ziml.areteem.org.

2.1 ZIML October 2016 Varsity

Below are the solutions from the Varsity ZIML Competition held in October 2016.

The problems from the contest are available on p.15.

Problem 1 Solution

If the largest number is 5, the two other numbers must be < 5 so chosen from $1, 2, 3, 4$. Hence $K = \binom{4}{2} = 6$.

Answer: 6

Problem 2 Solution

Let $a = xy, b = x + y$ then $ab = 884, a + b = 69$. By Vieta's Theorem, a and b are the roots of quadratic equation

$$t^2 - 69t + 884 = 0,$$

which is

$$(t - 17)(t - 52) = 0.$$

So either $a = 17, b = 52$, or $a = 52, b = 17$.

If $a = 17, b = 52$, then $xy = 17$ and $x + y = 52$, so by Vieta's Theorem, x and y are roots of the equation $u^2 - 52u + 17 = 0$, which has no positive integer roots.

If $a = 52, b = 17$, then $xy = 52$ and $x + y = 17$, so again by Vieta's Theorem, x and y are roots of the equation $u^2 - 17u + 52 = 0$, which factors to $(u - 4)(u - 13) = 0$, thus x and y are 4 and 13 in some order, and $x^2 + y^2 = 4^2 + 13^2 = 185$.

Answer: 185

Problem 3 Solution

If the total number is a power of 3, then the number 1 would be the one remaining. The largest power of 3 under 999 is 729, so we need to cross out $999 - 729 = 270$ numbers and find the next number in line.

At each step, we keep one number and cross out two, therefore we shall perform $270/2 = 135$ steps. Thus we skip 135 numbers and cross out 270, going past a total of 405 numbers. The next number is 406, and at this point there are 729 numbers remaining, so 406 will remain till the end.

Answer: 406

Problem 4 Solution

Pair up the first and last terms: since

$$\frac{200 \cdot 1}{71} + \frac{200 \cdot 70}{71} = 200,$$

we get

$$\left\lfloor \frac{200 \cdot 1}{71} \right\rfloor + \left\lfloor \frac{200 \cdot 70}{71} \right\rfloor = 200 - 1 = 199.$$

Similarly, pair up the second and second-to-last terms, we also get 199, and so on. There are $70/2 = 35$ pairs, thus the sum is $199 \cdot 35 = 6965$.

Answer: 6965

Problem 5 Solution

Let N be a magic number with k-digits ($k \geq 1$). Let x be any positive integer, then attaching N to the right of x will produce the new number $10^k x + N$. This should always be a multiple of N, so

$$10^k x + N = mN, \quad m \text{ is an integer.}$$

This means

$$10^k x = (m - 1)N$$

for all positive integer x. Let $x = 1$, then $10^k = (m-1)N$, so N must be a factor of 10^k.

Categorize according to the number of digits k of N:
Case (1): If $k = 1$, N has only one digit, and it should be a factor of 10. Thus $N = 1$, 2, or 5.
Case (2): If $k = 2$, N has two digits, and it should be a factor of 100, so $N = 10, 20, 25$, or 50.
Case (3): If $k = 3$, N has three digits, and it should be a factor of 1000, so $N = 100, 125, 200, 250, 500$.

Hence the magic numbers less than 1000 are: 1, 2, 5, 10, 20, 25, 50, 100, 125, 200, 250, 500. In total there are 12 of them.

Answer: 12

Problem 6 Solution
First the discriminant $m^2 - 16 \geq 0$, so $|m| \geq 4$.

Let x_1, x_2 be the two roots. By Vieta's formula, $x_1 x_2 = 4$, therefore at most one of them is in $[-1, 1]$, so according to the problem, exactly one root is in $[-1, 1]$. This means for the function $f(x) = x^2 - mx + 4$, $f(-1)$ and $f(1)$ must have different signs (or equal 0). So

$$f(-1)f(1) = (1 + m + 4)(1 - m + 4) \leq 0,$$

$$(m+5)(m-5) \geq 0,$$

which means $|m| \geq 5$. So for all m such that $|m| \geq 5$, at least one root of the equation is in $[-1, 1]$. Therefore, $L = 5$.

Answer: 5

Problem 7 Solution
There are $\binom{12}{5}$ ways to choose which 5 beads to use. There are $5!/5 = 4!$ ways to arrange them in a circle. Since we can get the

same arrangement either clockwise or counterclockwise (think of flipping the bracelet) we also have to divide by 2. Hence there are $\binom{12}{5} \cdot \frac{4!}{2} = 9504$ possible bracelets.

Answer: 9504

Problem 8 Solution

The second bird needs at most 3 tries to escape, but the first bird may need any number of tries. To calculate the probability that the second bird needs fewer tries than the first, we can calculate the opposite: the probability that the second bird needs the same number as or more tries than the first bird, and subtract that value from 1. There are 3 cases for this.

Case 1: The second bird escapes in only one try; the probability is $\frac{1}{3}$. So the first bird escapes in only one try too, also with probability $\frac{1}{3}$. The probability for this case is $\frac{1}{3} \times \frac{1}{3} = \frac{1}{9}$.

Case 2: The second bird escapes in two tries; the probability is still $\frac{1}{3}$. So the first bird escapes in one or two tries, with probability $\frac{1}{3} + \frac{2}{3} \times \frac{1}{3} = \frac{5}{9}$. The probability for this case is $\frac{1}{3} \times \frac{5}{9} = \frac{5}{27}$.

Case 3: The second bird escapes in three tries; the probability is still $\frac{1}{3}$. So the first bird escapes in one or two or three tries, with probability $\frac{1}{3} + \frac{2}{3} \times \frac{1}{3} + \frac{2}{3} \times \frac{2}{3} \times \frac{1}{3} = \frac{19}{27}$. The probability for this case is $\frac{1}{3} \times \frac{19}{27} = \frac{19}{81}$.

Combining the three cases, $\frac{1}{9} + \frac{5}{27} + \frac{19}{81} = \frac{43}{81}$, which is the probability that the second bird needs at least the same number of tries to escape as the first bird.

So the probability that the second bird needs fewer number of tries than the first bird is $1 - \frac{43}{81} = \frac{38}{81}$. Therefore $M = 38$ and $N = 81$, and the final answer (asking for M) is 38.

Answer: 38

Problem 9 Solution
Let $n^3 = 5p + 1$, then $5p = n^3 - 1 = (n-1)(n^2 + n + 1)$.

Since 5 and p are both primes, there are 4 possibilities: $n - 1 = 1$, $n - 1 = 5$, $n - 1 = p$, and $n - 1 = 5p$.

If $n - 1 = 1$, then $n = 2$, and $n^2 + n + 1 = 7$, and $(n-1)(n^2 + n + 1) = 7$ is not a multiple of 5. So $n - 1 = 1$ is not a solution.

If $n - 1 = 5$, $n = 6$, and $p = n^2 + n + 1 = 43$. This is a solution.

If $n - 1 = p$, then $n^2 + n + 1 = 5$. This equation does not give an integer solution.

If $n - 1 = 5p$, then $n^2 + n + 1 = 1$, we get $n = 0$ or $n = -1$, neither of which satisfies $n - 1 = 5p$ for some prime number p.

Therefore the only possibility is that $n = 6$ and $p = 43$. Thus the answer is 43.

Answer: 43

Problem 10 Solution
Let the radii of circles $\odot O_1, \odot O_2, \odot O_3$ be x, y, z respectively. Then $x - y = 3, x - z = 6, y + z = 7$. Solve and get $x = 8, y =$

$5, z = 2$, so the answer is 5.

Answer: 5

Problem 11 Solution

By Cauchy-Schwarz Inequality,

$$(a^2 + b^2 + c^2)(x^2 + y^2 + z^2) \geq (ax + by + cz)^2,$$

where equality occurs if and only if

$$\frac{a}{x} = \frac{b}{y} = \frac{c}{z}.$$

In this problem, equality actually occurs, so

$$\frac{a}{x} = \frac{b}{y} = \frac{c}{z} = \frac{5}{6},$$

therefore

$$\frac{a+b+c}{x+y+z} = \frac{5}{6},$$

and the final answer is $5 + 6 = 11$.

Answer: 11

Problem 12 Solution

As shown in the diagram,

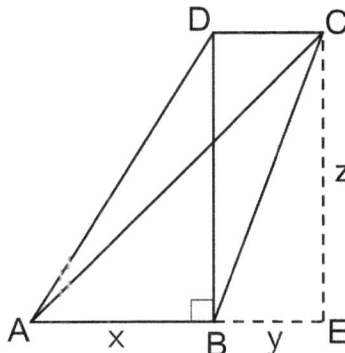

---

let x and y be the bases of the trapezoid, and z be the altitude, then

$$x+y+z = 16$$

$$\frac{1}{2}z(x+y) = 32$$

Solve and get $x+y=8$ and $z=8$. Through C construct $\overline{CE} \perp \overline{AB}$, where the foot E is on the extension of $\overline{AB}$. Then $AE = x+y = z = 8$, which means $\triangle ACE$ is an isosceles right triangle. Therefore $AC = \sqrt{AE^2 + CE^2} = 8\sqrt{2}$.

Answer: 128

Problem 13 Solution
The number 1 can only be on the numerator, and 2 can only be on the denominator. Everything else can be either on top or on the bottom: $2^6 = 64$. However, since $3 \times 8 = 4 \times 6$, the result is repeated when 3×8 is on top and 4×6 is on the bottom, and vice versa. In each of those cases there are $2^2 = 4$ choices to place 5 and 7, so these 4 choices should be subtracted from 64, getting 60 as the final answer.

Answer: 60

Problem 14 Solution
The solutions are all in the interval $(0, 32]$. The function $\sin 5\pi x$ has period $\frac{2}{5}$, and there are 80 periods in the interval $(0, 32]$. Within each period, there are two solutions (intersections of the graphs), except for the 3rd period which has only one. Thus the final answer is $2 \times 80 - 1 = 159$.

Answer: 159

Problem 15 Solution
$2^{10} = 1024 \equiv -1 \pmod{25}$, so $2^{20} \equiv 1 \pmod{25}$. Since $345 = 20 \times 17 + 5$, $2^{345} \equiv 2^5 \equiv 7 \pmod{25}$. It is clear that 2^{345} is a

multiple of 16. The only number under 400 that is a multiple of 16 and is 7 $\pmod{25}$ is 32.

Answer: 32

Problem 16 Solution

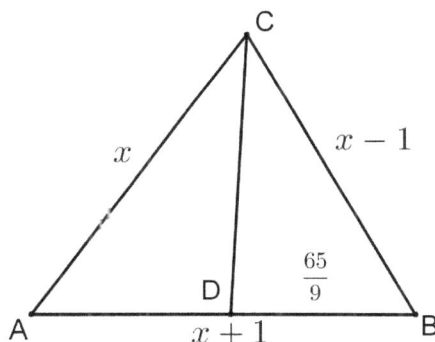

As shown in the diagram, let $AC = x$, then $BC = x - 1$, $AB = x + 1$. Let $\overline{CD}$ be the angle bisector, then $BD = \dfrac{65}{9}$. By Angle Bisector Theorem,

$$\frac{x}{x-1} = \frac{x + 1 - \dfrac{65}{9}}{\dfrac{65}{9}}.$$

Simplify to get $9x^2 - 130x + 56 = 0$, which is $(9x - 4)(x - 14) = 0$. So $x = \dfrac{4}{9}$ or $x = 14$. Since $x - 1 > 0$, the only valid solution is $x = 14$. So the side lengths are $13, 14, 15$.

Answer: 15

Problem 17 Solution

Draw $\overline{AH} \perp \overline{BC}$ at H. Let $x = CH$, then $AH = \sqrt{3}x$. Since $DH = AH, 8 + x = \sqrt{3}x$, thus $x = 4(\sqrt{3}+1)$. Since $BC = AC$, get $BD = BC - DC = AC - DC = 8\sqrt{3}$. Hence $A + B = 8 + 3 = 11$.

Answer: 11

Problem 18 Solution

Square the expression:

$$(|\sin x| + |\cos x|)^2 = \sin^2 x + \cos^2 x + 2|\sin x \cos x| = 1 + |\sin 2x|.$$

Since $|\sin 2x| \leq 1$, $(|\sin x| + |\cos x|)^2 \leq 2$, therefore the maximum value is $\sqrt{2}$, and $K = 2$.

Answer: 2

Problem 19 Solution

Place one Asimov book at each the of (A) location and one Clarke book at each of the (C) location. Then there are 3 extra Asimov books and 4 extra Clarke books. To count the ways to place these extra books, we use Stars and Bars for each of the authors.

For Asimov books, $a + b + c + d = 3$, thus there are $\binom{3+3}{3} = 20$ ways.

For Clarke books, $x + y + z + w = 4$, thus there are $\binom{4+3}{4} = 35$ ways.

By the product rule, the final answer is $20 \times 35 = 700$.

Answer: 700

Problem 20 Solution

Reflect D over $\overline{AB}$ to get D_1 and over $\overline{AC}$ to get D_2. Then $BD = BD_1 = 2$, $CD = CD_2 = 3$, $\angle D_1 = \angle D_2 = 90°$. Since angle $A = 45°$, $\angle D_1AD_2 = 90°$, so we can extend D_1B and D_2C to meet at

E and form a square as in the diagram below.

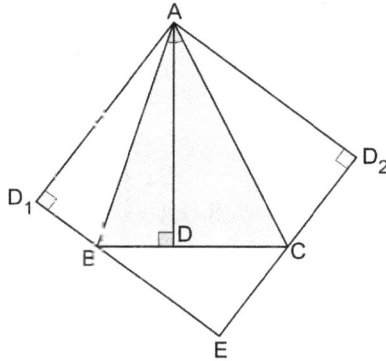

Call $AD = h$, so in fact the square has side length h. As $BE = h - BD_1 = h - 2$ and $CE = h - CD_2 = h - 3$, the Pythagorean theorem gives us $5^2 = (h-2)^2 + (h-3)^2$. Solving for h (and removing any negative solutions) we get $h = 6$. Hence $[ABC] = 5 \cdot 6/2 = 15$.

Answer: 15

2.2 ZIML November 2016 Varsity

Below are the solutions from the Varsity ZIML Competition held in November 2016.
The problems from the contest are available on p.21.

Problem 1 Solution

Squaring we have $x - 4 = x^2 - 12x + 36$ so $x^2 - 13x + 40 = (x - 8)(x - 5) = 0$. Hence $x = 5, 8$ but $x = 5$ is extraneous.

Answer: 8

Problem 2 Solution

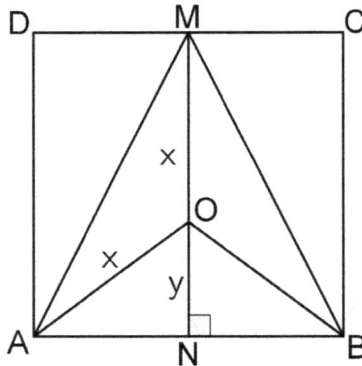

As shown in the diagram, let $OA = OB = OC = x$, and $ON = y$, then $x + y = 6$ and $x^2 - y^2 = 3^2$. Solve this system to get $x = \frac{15}{4}, y = \frac{9}{4}$. Let r be the inradius of $\triangle OAB$, and $2l$ be the perimeter of $\triangle OAB$, then

$$[OAB] = \frac{1}{2} \cdot 2l \cdot r = lr,$$

so

$$r = \frac{[OAB]}{l} = \frac{AB \cdot ON/2}{(AB + 2x)/2} = \frac{6 \cdot \frac{9}{4}}{6 + 2 \cdot \frac{15}{4}} = 1.$$

Therefore the diameter of the incircle of $\triangle OAB$ is 2.

Answer: 2

Problem 3 Solution

There are $2^{10} = 1024$ total outcomes. There are $\binom{10}{5} = 252$ ways to get an equal amount of heads and tails. By symmetry, half of the remaining outcomes have more heads than tails, hence there are 386 outcomes with more heads than tails. The ratio is thus $\frac{386}{1024} = \frac{193}{512}$ so $p + q = 705$.

Answer: 705

Problem 4 Solution

The number between 1 and 2000 with the highest power of 2 in its prime factorization is $2^{10} = 1024$. Since N is the least common multiple of 1, 2, 3, ..., 1998, 1999, 2000, the power of 2 in the prime factorization of N is 10.

Answer: 10

Problem 5 Solution

Using the change of base formula we can rewrite the equation as

$$2\log_3 x + \log_3 x + \frac{1}{2}\log_3 x = -\frac{21}{2} \text{ so } \log_3 x = -3.$$

Hence $x = 3^{-3} = 1/27$.

Answer: 28

Problem 6 Solution

The centers of two ground balls on the diagonal of the square are $20\sqrt{2}$cm apart, as shown in the diagram.

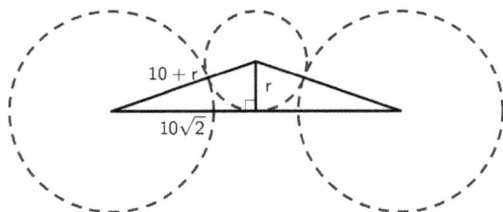

Let r be the radius of the fifth ball, then we have equation

$$(10\sqrt{2})^2 + r^2 = (10+r)^2,$$

solve and get $r = 5$ centimeters.

Answer: 5

Problem 7 Solution

Each time Ed takes a ball out of the bag, the probability of getting a red ball is $\frac{5}{9}$, and the probability of getting a white ball is $\frac{4}{9}$. He can get a red ball twice and a white ball once in three different ways: RRW, RWR, and WRR. Therefore the overall probability is

$$3 \times \frac{5}{9} \times \frac{5}{9} \times \frac{4}{9} = \frac{300}{729} = \frac{100}{243}.$$

and the final answer is $100 + 243 = 343$.

Answer: 343

Problem 8 Solution

First, note that $40500 = 2^2 3^4 5^3$. Let $a = 2^{x_2} 3^{x_3} 5^{x_5}$, $b = 2^{y_2} 3^{y_3} 5^{y_5}$, $c = 2^{z_2} 3^{z_3} 5^{z_5}$, where all the exponents are nonnegative integers, then

$$x_2 + y_2 + z_2 = 2,$$
$$x_3 + y_3 + z_3 = 4,$$
$$x_5 + y_5 + z_5 = 3.$$

These three equations represent three 'stars and bars' problems, so the answer is

$$\binom{2+3-1}{2}\binom{4+3-1}{4}\binom{3+3-1}{3} = 900.$$

Answer: 900

Problem 9 Solution

$2^{10} = 1024 \equiv -1 \pmod{25}$, so $2^{20} \equiv 1 \pmod{25}$. Since $987 = 20 \times 49 + 7$, $2^{987} \equiv 2^7 \equiv 3 \pmod{25}$. It is clear that 2^{987} is a multiple of 4. There is only one multiple of 4 under 100 that is also 3 $\pmod{25}$: it is 28.

Answer: 28

Problem 10 Solution

Multiply both sides of the equation by 2, and we get

$$px^2 - qx + 2017 \times 2 = 0.$$

Let x_1 and x_2 be the two roots, which are both prime numbers. By Vieta's formulas,

$$x_1 \cdot x_2 = \frac{2017 \times 2}{p},$$

and since x_1 and x_2 are primes, the only possibility is that they are 2017 and 2 (note that 2017 is prime) and $p = 1$.

Also by Vieta's,

$$x_1 + x_2 = \frac{q}{p} = q,$$

therefore $q = 2017 + 2 = 2019$.

Answer: 2019

Problem 11 Solution

The vertical line $x = 1$ is one such line. If the line is not vertical, let k be its slope, then the equation of the line is $y = k(x-1)$. The common points of the line and the parabola satisfy the following equation:

$$k(x-1) = \frac{x^2}{2},$$

which is

$$\frac{x^2}{2} - kx + k = 0.$$

Since there is exactly one common point, the discriminant of the quadratic equation should be 0, thus

$$k^2 - 2k = 0,$$

and therefore $k = 0$ or $k = 2$. The corresponding lines are $y = 0$ and $y = 2x - 2$. Including the line $x = 1$, there are 3 lines that have exactly one common point with the parabola.

Answer: 3

Problem 12 Solution

First we get $n \geq 4$, otherwise 2 must be adjacent to 1 or 3. With no restrictions there are $n!$ numbers. Now use PIE (Principle of Inclusion-Exclusion): If 2 and 1 are adjacent, bundle them together, $(n-1)!$ ways; and 2 and 1 can be switched, so $2 \cdot (n-1)!$ ways. If 2 and 3 are adjacent, similarly $2 \cdot (n-1)!$ ways. If 2 is adjacent to both 1 and 3, $2 \cdot (n-2)!$ ways. Finally,

$$n! - 2 \cdot 2 \cdot (n-1)! + 2 \cdot (n-2)! = 2400.$$

Simplify, $(n-2)!(n-2)(n-3) = 2400$, thus $n = 7$ by trial and error.

Answer: 7

Problem 13 Solution

Since a, b, c, d are roots of $x^4 - 5x^2 + 3x + 1 = 0$ or $x^4 = 5x^2 - 3x - 1$ we can write

$$a^4 + b^4 + c^4 + d^4 = 5(a^2 + b^2 + c^2 + d^2) - 3(a+b+c+d) - 4.$$

We know $a + b + c + d = 0$ by Vieta's Formulas. Further we also know

$$
\begin{aligned}
& a^2 + b^2 + c^2 + d^2 \\
= {} & (a+b+c+d)^2 - 2(ab + ac + ad + bc + bd + cd) \\
= {} & 0 - 2(-5) \\
= {} & 10.
\end{aligned}
$$

Hence

$$
\begin{aligned}
& a^4 + b^4 + c^4 + d^4 \\
= {} & 5(a^2 + b^2 + c^2 + d^2) - 3(a+b+c+d) - 4 \\
= {} & 5 \cdot 10 - 3 \cdot 0 - 4 \\
= {} & 46.
\end{aligned}
$$

Answer: 46

Problem 14 Solution

$210 = 2 \times 3 \times 5 \times 7$. Let $m = 2^a 3^b 5^c 7^d$, and $n = 2^p 3^q 5^r 7^s$. The exponents a and p must satisfy $\max(a, p) = 1$, so there are 3 possibilities: $a = 0, p = 1$; $a = 1, p = 0$; $a = 1, p = 1$. Similarly, each of the pairs of exponents, b and q, c and r, d and s, has 3 possibilities. Therefore the total number of possible ordered pairs (m, n) is $3 \times 3 \times 3 \times 3 = 81$.

Answer: 81

Problem 15 Solution

Since $10000 = 2^4 \times 5^4$, its even positive divisors are of the form $2^a 5^b$ where $1 \leq a \leq 4$ and $0 \leq b \leq 4$, therefore the sum of these even divisors is

$$(2 + 2^2 + 2^3 + 2^4)(1 + 5 + 5^2 + 5^3 + 5^4) = 23430.$$

Answer: 23430

Problem 16 Solution

We have that

$$
\begin{aligned}
\frac{1}{\cos 2\alpha} + \tan 2\alpha &= \frac{1}{\cos 2\alpha} + \frac{\sin 2\alpha}{\cos 2\alpha} \\
&= \frac{1 + \sin 2\alpha}{\cos 2\alpha} \\
&= \frac{\sin^2 \alpha + \cos^2 \alpha + 2\sin\alpha\cos\alpha}{\cos 2\alpha} \\
&= \frac{(\cos\alpha + \sin\alpha)^2}{\cos^2\alpha - \sin^2\alpha} \\
&= \frac{\cos\alpha + \sin\alpha}{\cos\alpha - \sin\alpha} \\
&= \frac{1 + \tan\alpha}{1 - \tan\alpha}
\end{aligned}
$$

so the answer is 2017.

Answer: 2017

Problem 17 Solution

Subtract 1 from each fraction, then

$$\frac{x-2}{x+2} - 1 + \frac{x-8}{x+8} - 1 = \frac{x-4}{x+4} - 1 + \frac{x-6}{x+6} - 1,$$

$$\frac{-4}{x+2} + \frac{-16}{x+8} = \frac{-8}{x+4} + \frac{-12}{x+6},$$

$$\frac{1}{x+2} + \frac{4}{x+8} = \frac{2}{x+4} + \frac{3}{x+6},$$

$$\frac{4}{x+8} - \frac{3}{x+6} = \frac{2}{x+4} - \frac{1}{x+2},$$

$$\frac{4(x+6) - 3(x+8)}{(x+8)(x+6)} = \frac{2(x+2) - (x+4)}{(x+4)(x+2)},$$

$$\frac{x}{(x+8)(x+6)} = \frac{x}{(x+4)(x+2)}.$$

Since $x \neq 0$, we get $(x+8)(x+6) = (x+4)(x+2)$, which simplifies to $14x + 48 = 6x + 8$, so $8x = -40$, and then $x = -5$.

Answer: -5

Problem 18 Solution
One-to-one correspondence: $a' = a, b' = b - 2, c' = c - 4$, then select a', b', c' from $\{1, 2, \ldots, 10\}$, there are $\binom{10}{3} = 120$ ways.

Answer: 120

Problem 19 Solution
Connect and extend $\overline{CO_1}$, and $\overline{BO_3}$, intersecting at I, as shown. Also connect $\overline{O_1O_3}$, and construct $\overline{ID} \perp \overline{BC}$ at D. Let P be the intersection of $\overline{ID}$ and $\overline{O_1O_3}$.

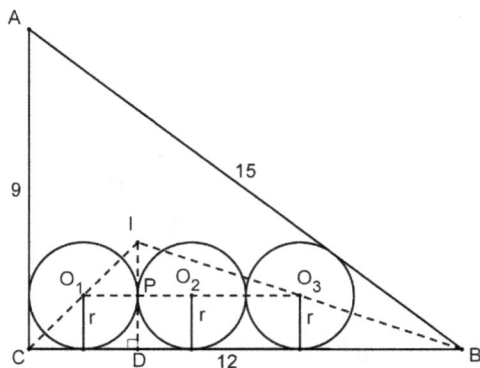

Clearly $\overline{CO_1}$ and $\overline{BO_3}$ bisect $\angle ACB$ and $\angle ABC$ respectively, so I is the incenter of $\triangle ABC$, and ID is the inradius.

Since $(9, 12, 15)$ is a Pythagorean triple, $\triangle ABC$ is a right triangle with $\angle ACB = 90°$. Thus the area of $\triangle ABC$ is $\frac{1}{2} \times 12 \times 9 = 54$. Also the semiperimeter of $\triangle ABC$ is $\frac{9 + 12 + 15}{2} = 18$, therefore the inradius $ID = \frac{54}{18} = 3$. Note that $\overline{O_1 O_3} \parallel \overline{BC}$, hence $\triangle IO_1 O_3 \sim \triangle ICB$. $\overline{IP}$ and $\overline{ID}$ are the altitudes of $\triangle IO_1 O_3$ and $\triangle ICB$ respectively, thus $\frac{IP}{ID} = \frac{O_1 O_3}{CB}$. We have $ID = 3$, $IP = 3 - r$, $CB = 12$, and $O_1 O_3 = 4r$, so we have equation

$$\frac{3 - r}{3} = \frac{4r}{12},$$

solving for r, we get $r = 1.5$.

Answer: 1.5

Problem 20 Solution

If two real numbers x and y such that $x - y > 1$, we must have $\lfloor x \rfloor \neq \lfloor y \rfloor$. So if k is large enough, $\left\lfloor \dfrac{k^2}{2000} \right\rfloor$ and the terms after it are all distinct. The values in the first $k - 1$ terms can be calculated separately. In order to find such k, we solve the inequality $\dfrac{(k+1)^2}{2000} - \dfrac{k^2}{2000} > 1$, and get $k > 999.5$. Thus from the 1000th terms through the 2000th term, the values of $\left\lfloor \dfrac{1000^2}{2000} \right\rfloor, \left\lfloor \dfrac{1001^2}{2000} \right\rfloor,$ $\ldots, \left\lfloor \dfrac{2000^2}{2000} \right\rfloor$ are all distinct, and in total there are 1001 values.

Now we consider the cases where $k \le 999$.

If $k \le 44$, $0 < \dfrac{k^2}{2000} < 1$, thus the 44 numbers $\left\lfloor \dfrac{1^2}{2000} \right\rfloor$ through $\left\lfloor \dfrac{44^2}{2000} \right\rfloor$ are all 0.

If $45 \le k \le 63$, $1 < \dfrac{k^2}{2000} < 2$, thus $\left\lfloor \dfrac{k^2}{2000} \right\rfloor$ are all 1. And so on.

Since $\left\lfloor \dfrac{998^2}{2000} \right\rfloor = 498$, and $\left\lfloor \dfrac{999^2}{2000} \right\rfloor = 499$, the values 0, 1, ..., 499 all appear in the sequence. Therefore there are $1001 + 500 = 1501$ distinct values.

Answer: 1501

2.3   ZIML December 2016 Varsity

Below are the solutions from the Varsity ZIML Competition held in December 2016.

The problems from the contest are available on p.27.

Problem 1 Solution

Since $n^2 + 1 = n^2 - 1 + 2 = (n+1)(n-1) + 2$, we have $(n+1) \mid 2$. So $n = 1$ is the only such positive integer.

Answer: 1

Problem 2 Solution

First arrange the 5 couples in a circle, which can be done in $5!/5 = 4! = 24$ ways. Then each couple themselves has $2! = 2$ arrangements, so our final answer is

$$24 \cdot 2^5 = 738.$$

Answer: 768

Problem 3 Solution

Use the Law of Cosines to get $3^2 = 3^2 + 4^2 - 2 \cdot 3 \cdot 4 \cos(\angle CBD)$ and solve to get $\cos(\angle CBD) = \dfrac{2}{3}$. Hence using the Law of Cosines again we have $(AD)^2 = 9^2 + 4^2 - 2 \cdot 9 \cdot 4 \cdot \cos(\angle CBD) = 81 + 16 - 72 \cdot \dfrac{2}{3} = 49$ so $AD = 7$.

Answer: 7

Problem 4 Solution

We have $\log_{xy} z = 2$ so $\log_z(xy) = \log_z x + \log_z y = 1/2$. We also have $\log_z x - \log_z y = 1$. Solving this system we get $\log_z y = -1/4$, so $\log_y z = -4$.

Answer: -4

Problem 5 Solution

Let $N = \overline{abcde}$, where a,b,c,d,e are distinct nonzero digits.
The number of 3-digit numbers made up of 3 of the 5 digits of N
is $5 \times 4 \times 3 = 60$. Among these numbers, each digit of a,b,c,d,e
appears exactly the same number of times at each place, which
is 12 times. Therefore the sum of all the 3-digit numbers is
$(a+b+c+d+e) \times 12 \times 111 = (a+b+c+d+e) \times 1332$, which,
according to the question, is also equal to N.
Since 1332 is a multiple of 9, so is N, thus $9 \mid (a+b+c+d+e)$.
Also, $a+b+c+d+e$ is at least $1+2+3+4+5 = 15$ and at
most $5+6+7+8+9 = 35$, so it must be either 18 or 27. Only
27 satisfies the requirements, and we get $N = 27 \times 1332 = 35964$.

Answer: 35964

Problem 6 Solution

Consider the side view drawn below (where r is the radius of the
sphere).

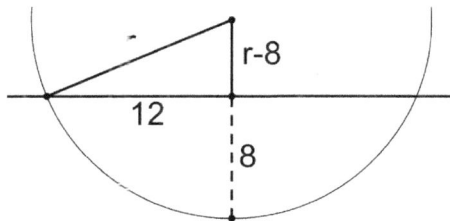

The triangle shown is a right triangle, so

$$r^2 = 12^2 + (r-8)^2.$$

Solving for r gives $r = 13$.

Answer: 13

Problem 7 Solution

$$\begin{aligned}
&\frac{a}{b+c} + \frac{b}{c+a} + \frac{c}{a+b} \\
&= (a+b+c)\left(\frac{1}{b+c} + \frac{1}{c+a} + \frac{1}{a+b}\right) - 3 \\
&\geq \frac{1}{2}(1+1+1)^2 - 3 \\
&= \frac{3}{2}
\end{aligned}$$

using the Cauchy-Schwarz Inequality.

Answer: 1.5

Problem 8 Solution
Using patterns or Fermat's Little Theorem, $10^6 \equiv 1 \pmod 7$. Hence

$$10^{10} \equiv 10^{10^2} \equiv \cdots \equiv 10^{10^{10}} \equiv 10^4 \equiv 3^4 \equiv 4 \pmod 7.$$

Thus the sum is equivalent to

$$10 \cdot 4 \equiv 5 \pmod 7$$

so the remainder is 5.

Answer: 5

Problem 9 Solution
The probability that X is even is

$$\begin{aligned}
&0.6 \cdot 0.4 + 0.6^3 \cdot 0.4 + 0.6^5 \cdot 0.4 + \cdots \\
&= 0.6 \cdot 0.4 \cdot (1 + 0.36^2 + 0.36^4 + \cdots).
\end{aligned}$$

Using a geometric series we have this is equal to

$$\frac{0.24}{0.64} = \frac{3}{8} = 37.5\%,$$

so $K = 37.5$.

Answer: 37.5

Problem 10 Solution
Since $z^7 = 1$ and $z \neq 1$, we have

$$1 + z + z^2 + z^3 + z^4 + z^5 + z^6 = 0.$$

Then

$$
\begin{aligned}
\frac{z}{1+z^2} + \frac{z^2}{1+z^4} &= \frac{1}{z^6+z} + \frac{1}{z^5+z^2} \\
&= \frac{z+z^2+z^5+z^6}{z^3(1+z^3)(1+z^5)} \\
&= \frac{z(1+z)(1+z^4)}{z^3(1+z^3)(1+z^5)} \\
&= \frac{z^2(1+z)(1+z^4)}{z^4(1+z^3)(1+z^5)} \\
&= \frac{z^2(1+z)(1+z^4)}{(z^4+1)(1+z^5)} \\
&= \frac{z^2+z^3}{1+z^5},
\end{aligned}
$$

and then

$$\frac{z}{1+z^2}+\frac{z^2}{1+z^4}+\frac{z^3}{1+z^6} = \frac{z^2+z^3}{1+z^5}+\frac{z^3}{1+z^6}$$

$$= \frac{z^2+z^8+z^3+z^9+z^3+z^8}{1+z^5+z^6+z^{11}}$$

$$= \frac{z^2+z+z^3+z^2+z^3+z}{1+z^4+z^5+z^6}$$

$$= \frac{2(z+z^2+z^3)}{-(z+z^2+z^3)}$$

$$= -2.$$

Answer: -2

Problem 11 Solution
First note $\sin C = \dfrac{5}{13}$ as $(5,12,13)$ is a Pythagorean Triple. Similarly, $\sin B = \dfrac{7}{25}$. Recall angles in a triangle add up to π radians. Thus we have

$$\sin A = \sin(\pi - B - C) = \sin(B+C)$$

$$= \sin B \cos C + \cos B \sin B = \frac{24}{25}\cdot\frac{5}{13}+\frac{7}{25}\cdot\frac{12}{13}$$

$$= \frac{204}{325}.$$

Hence $Q - P = 121$.

Answer: 121

Problem 12 Solution
Consider the times Jack and Jill arrive in a 4×4 square (so 1 unit is 15 minutes). Taking into account how long each will wait (or

not wait) we get that Jack and Jill will have dinner together if they arrive within the shaded region given below.

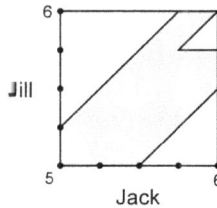

Note the shaded region is the entire square minus three triangles (with area $9/2, 1/2, 2$ respectively). Hence the probability is

$$\frac{16 - 9/2 - 1/2 - 2}{16} = \frac{9}{16}.$$

Answer: 25

Problem 13 Solution

Consider the binomial expansion of $(1+x)^{2016}$, and let $x = i$ and $x = -i$, where $i = \sqrt{-1}$, and add to get the sum above is equal to

$$\frac{(1+i)^{2016} + (1-i)^{2016}}{2} = \frac{2^{1008}(e^{i504\pi} + e^{-i504\pi})}{2}$$

$$= \frac{2^{1009}}{2} = 2^{1008}.$$

Hence the largest (and only) exponent is 1008.

Answer: 1008

Problem 14 Solution

Let $x = AB$. Then as $\triangle ABD$ is a 45-45-90 triangle, so $DB = x$ as well. As $\triangle ABC$ is a 30-60-90 triangle, $BC = \sqrt{3}AB$ so $x + 10 =$

$x\sqrt{3}$. Hence

$$x = \frac{10}{\sqrt{3}-1} = 5(\sqrt{3}+1).$$

Using $\sqrt{3} \approx 1.73$ we have that $AB \approx 13.65$ so AB rounded to the nearest integer is 14.

Answer: 14

Problem 15 Solution

Note $x^2 + x - 2 + 2x^2 - 5x + 3 = 3x^2 - 4x + 1$ so this is $A^2 + B^2 = (A+B)^2$ and hence $AB = 0$. Therefore we have

$$(x^2 + x - 2)(2x^2 - 5x + 3) = 0.$$

The first quadratic has roots $1, -2$ and the second quadratic has roots $1, 3/2$. Thus there are 3 distinct real roots.

Answer: 3

Problem 16 Solution

By Fermat's Little Theorem, $4^{10} \equiv 1 \pmod{11}$, so we can examine only $a_1 = 0, 1, 2, \ldots, 9$. Further, note that $4^3 \equiv 9 \pmod{11}$ and $4^9 \equiv 3 \pmod{11}$, so if the sequence ever reaches either $3, 9$ the rest of the sequence alternates between $3, 9$ forever. It is routine to check that in fact, every such sequence reaches 3 or 9 within the first 5 terms. Further, a_5 can be either $3, 9$ which implies that a_{100} is either 3 or 9 so the sum of all possibilities is 12.

Answer: 12

Problem 17 Solution

Let $CD = x$. Using power of a point we have that

$$DA^2 = DC \cdot DB \Rightarrow (10\sqrt{6})^2 = x(x+10) \Rightarrow 600 = x(x+10),$$

so we can solve to get $x = 20$ (or $x = -30$ which is impossible). Using the Pythagorean theorem, we get $DN = \sqrt{20^2 - 10^2} =$

$\sqrt{300} = 10\sqrt{3}$. Then as $\triangle CDN \sim \triangle BDM$ we get that

$$BM = \frac{BD}{2} = 15, DM = 15\sqrt{3}.$$

Hence

$$MN = 15\sqrt{3} - 10\sqrt{3} = 5\sqrt{3}.$$

Thus $MBCN$ is a trapezoid with height MN and bases BM and CN, so has area

$$\frac{1}{2}(15 + 10) \cdot 5\sqrt{3} = 125\sqrt{3}.$$

Answer: 128

Problem 18 Solution

By Wilson's Theorem, $31 \mid 30! + 1$. Now $30! + 1 = 30 \cdot 29! + 1 \equiv (-1) \cdot 29! + 1 \equiv -(29! - 1) \pmod{31}$. Hence $31 \mid 29! - 1$. If $p < 31$ is prime, then $p \mid 29!$, so $p \nmid 29! - 1$, hence 31 is the smallest prime factor of $29! - 1$.

Answer: 31

Problem 19 Solution

We know $a + b + c + d = 0$ so $a + b = -(c + d)$. Cubing both sides gives us that

$$a^3 + 3a^2 b + 3ab^2 + b^3 = -(c^3 + 3c^2 d + 3cd^2 + d^3).$$

Thus,

$$a^3 + b^3 + 3ab(a + b) = -(c^3 + d^3) - 3cd(c + d),$$

and so

$$a^3 + b^3 + c^3 + d^3 = 3ab(c + d) + 3cd(a + b)$$
$$= 3(abc + abd + cda + cdb).$$

Therefore

$$\frac{a^3 + b^3 + c^3 + d^3}{abc + bcd + cda + dab} = \frac{3(abc + bcd + cda + dab)}{abc + bcd + cda + dab} = 3.$$

Answer: 3

Problem 20 Solution

Using stars and bars there are

$$\binom{201 + 3 - 1}{201} = 20503$$

ways if we are allowed to repeat numbers. There is 1 way ($201 = 67 + 67 + 67$) for 201 to be written as the sum of one repeated number. Now consider having a pair of repeated numbers. That is we have $a + a + b = 2a + b = 201$ for non-negative a, b. We can list the outcomes as ordered pairs (a, b):

$$(0, 201), (1, 199), (2, 197), \ldots (67, 67), \ldots, (100, 1),$$

that is, there are $101 - 1 = 100$ new possibilities (as we already dealt with $201 = 67 + 67 + 67$ above). As each of these can be arranged in 3 different ways, there are a total of 300 possibilities for a pair of repeated numbers. Subtracting from the total outcomes gives a final answer of

$$20503 - 300 - 1 = 20202.$$

Answer: 20202

2.4 ZIML January 2017 Varsity

Below are the solutions from the Varsity ZIML Competition held in January 2017.
The problems from the contest are available on p.33.

Problem 1 Solution

Note $68^3 \equiv 3^3 \equiv 1 \pmod{13}$ and $77 \equiv -1 \pmod{13}$. Hence

$$68^{50} + 77^{65} \equiv 3^2 + (-1) \equiv 8 \pmod{13}.$$

Answer: 8

Problem 2 Solution

Using the Law of Cosines we have

$$BC^2 = 9 + 8 - 12\sqrt{2}\cos 45° = 5$$

so $BC = \sqrt{5}$. Similarly, using the Law of Cosines again we can calculate

$$\cos B = \frac{\sqrt{5}}{5}.$$

Since $\sin^2 B + \cos^2 B = 1$ we get

$$\sin B = \frac{2\sqrt{5}}{5},$$

and hence

$$\tan B = \frac{\sin B}{\cos B} = 2.$$

Answer: 2

Problem 3 Solution

We have

$$(x^2 + x + 1) + (x^2 + 2x + 3) + (x^2 + 3x + 5) + \cdots$$
$$+ (x^2 + 25x + 49) = 6375$$

is equivalent to

$$25x^2 + 325x + 625 = 6375 \text{ or } x^2 + 13x - 230 = 0.$$

Factoring we get $x = 10$ or $x = -23$, so the negative solution is -23.

Answer: -23

Problem 4 Solution

There are 8! ways of placing 8 mutually non-attacking rooks on the board (one in each row but you can choose which columns). Then we have 8 choices for which rook to remove so that 7 remain. This gives a total of

$$8 \cdot 8! = 322560$$

arrangements. Note: This works because for any 7 rooks on the chessboard there is only one place you could put an 8th rook.

Answer: 322560

Problem 5 Solution

Any number and the sum of its digits are congruent modulo 9. Thus their difference must be a multiple of 9. Thus $3 + 1 + d + 2 = d + 6$ must be a multiple of 9, so $d = 3$. One such number is 3140.

Answer: 3

Problem 6 Solution

Draw a line parallel to $\overline{BC}$ through A, intersecting $\overline{CD}$ at E. Then form rectangle $BCFG$ as in the diagram below.

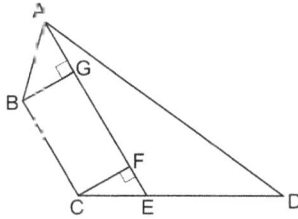

Since $\angle ABC = 135°$, $\triangle ABG$ is a 45-45-90 triangle and $AG = BG = \sqrt{3}$. As $BCFG$ is a rectangle, $F = 5 - \sqrt{3}$ and $CF = \sqrt{3}$. Similar reasoning ($\triangle CEF$ is a 30-60-90 triangle) gives $FE = 1$ and $CE = 2$. Note $AE = 6$ and $DE = 4$ so using the Law of Cosines on $\triangle AED$ gives

$$AD^2 = 6^2 + 4^2 - 2 \cdot 6 \cdot 4 \cos 120° = 36 + 16 + 24 = 76.$$

Answer: 76

Problem 7 Solution

First note that 37 is one such x. Since $\text{lcm}(11, 17) = 187$, all the solutions are of the form

$$x = 37 + 187k$$

for integers k. We have

$$37 + 187 \times (-6) = -1085 < -1000$$

and

$$37 + 187 \times 6 = 1159 > 1000,$$

so we want solutions $x = 37 + 187k$ for integers $k = -5, -4, \cdots$, $4, 5$. By symmetry we have

$$\sum_{k=-5}^{5} 37 + 187k = 11 \times 37 = 407,$$

as our answer.

Answer: 407

Problem 8 Solution
Reflect $\triangle ECD$ across $\overline{DE}$, so that point C is reflected to C'.
$\angle C' = \angle ECD = \angle B = \angle BCA$, so $\overline{DC'} \parallel \overline{BC}$. Extend $\overline{BC}$ to F
so that $\overline{DF} \parallel \overline{CC'}$, then $\angle F = \angle BCA = \angle B$, thus $BD = DF$. Also
$CFDC'$ is a parallelogram, thus $BD = DF = CC' = 2CE$ so $BD = 2 \cdot 3 = 6$.

Answer: 6

Problem 9 Solution
By Viete's formulas, $a+b+c = 5, ab+bc+ca = 7, abc = -1$.
Note that

$$
\begin{aligned}
(a+b+c)^3 =& a^3 + b^3 + c^3 + 3a^2b + 3ab^2 + 3b^2c + 3bc^2 \\
& + 3c^2a + 3ca^2 + 6abc \\
=& a^3 + b^3 + c^3 + 3(a+b+c)(ab+bc+ca) \\
& - 3abc.
\end{aligned}
$$

Hence

$$
a^3 + b^3 + c^3 = (5)^3 - 3(5)(7) - 3(-1) = 23.
$$

Answer: 23

Problem 10 Solution
There are $4^4 = 256$ total outcomes, so we just need to calculate
how many ways exactly one box can be empty. There are 4
choices for which box is empty and then 3 choices for which box
gets an extra ball. We then choose which two balls are in that

box, and arrange the remaining two balls. Hence there are

$$4 \cdot 3 \cdot \binom{4}{2} \cdot 2! = \binom{4}{2} \cdot 4! = 144$$

ways of getting exactly one empty box. Hence $P = 144$.

Answer: 144

Problem 11 Solution

Construct $\overline{CE} \parallel \overline{DB}$ intersecting the extension of $\overline{AB}$ at E. Then $CE = AC = 2$, $[ACE] = [ABCD] = \sqrt{2}$ and $\angle ACE = \angle AOB$. Also, $[ACE] = \frac{1}{2} \cdot 2 \cdot 2 \cdot \sin\angle ACE$, thus $\sin\angle ACE = \frac{\sqrt{2}}{2}$. Therefore $\angle ACE = 45°$, and then $\angle AOB = 45°$.

Answer: 45

Problem 12 Solution

Jane has a $\frac{1}{6}$ chance to win on any roll. Since the game stops at the first 6, to win on the kth roll, none of the other rolls can be a 6. hence the probability Jane wins is

$$\frac{1}{6} + \left(\frac{5}{6}\right)^2 \cdot \frac{1}{6} + \left(\frac{5}{6}\right)^4 \cdot \frac{1}{6} + \left(\frac{5}{6}\right)^6 \cdot \frac{1}{6} + \cdots$$

Using a geometric series we have the probability is

$$\frac{1}{6}\left[1 + \left(\frac{25}{36}\right) + \left(\frac{25}{35}\right)^2 + \left(\frac{25}{36}\right)^3 + \cdots\right] = \frac{1}{6} \cdot \frac{36}{11} = \frac{6}{11}$$

so $P + Q = 6 + 11 = 17$.

Answer: 17

Problem 13 Solution

First assume $a,b,c > 0$. Note that $40500 = 2^2 3^4 5^3$. Think of putting two 2's into three boxes (labeled a,b,c), then four 3's, etc.

By stars and bars, there are thus

$$\binom{2+3-1}{2}\binom{4+3-1}{4}\binom{3+3-1}{3} = 900$$

solutions for positive a, b, c. In addition, for any of these positive solutions, there are 3 ways to choose a pair of minus signs so that $a \cdot b \cdot c = (-a) \cdot (-b) \cdot c$, etc. Hence there are

$$4 \cdot 900 = 3600$$

solutions in total.

Answer: 3600

Problem 14 Solution
For all real numbers x,

$$x^2 - x + 1 = x^2 - x + \frac{1}{4} + \frac{3}{4} = \left(x - \frac{1}{2}\right)^2 + \frac{3}{4} > 0,$$

so the inequality is equivalent to

$$|3x^2 + kx + 8| \le 8(x^2 - x + 1),$$

which means

$$-8(x^2 - x + 1) \le 3x^2 + kx + 8 \le 8(x^2 - x + 1).$$

Solve one side first:

$$3x^2 + kx + 8 \le 8x^2 - 8x + 8,$$

hence

$$5x^2 - (8+k)x \ge 0$$

holds for all $x \in \mathbb{R}$. This means the discriminant $(8+k)^2 \le 0$, and the only way this can happen is that $k = -8$.

Now it remains to verify that

$$-8(x^2 - x + 1) \le 3x^2 - 8x + 8$$

holds true for all $x \in \mathbb{R}$ This is equivalent to $11x^2 - 16x + 8 \ge 0$ for all $x \in \mathbb{R}$, which can be verified by making sure the discriminant is negative: $16^2 - 4 \times 11 \times 16 < 0$. Therefore $k = -8$ works out fine.

So there is only one possible value for k, which is -8.

Answer: -8

Problem 15 Solution
This is a quartic in x, but if we set $y = 3$ we have $-2x^4 + 2x^3 - x^2 y - xy + y^2 = y^2 - (x^2 + x)y + (2x^3 - 2x^4) = 0$ which is quadratic in y. Using the quadratic formula we get

$$y = \frac{1}{2}\left(x^2 + x \pm \sqrt{(x^2+x)^2 - 4 \cdot (2x^3 - 2x^4)}\right)$$
$$= \frac{1}{2}\left(x^2 + x \pm \sqrt{x^4 + 2x^3 + x^2 - 8x^3 + 8x^4}\right)$$
$$= \frac{1}{2}\left(x^2 + x \pm \sqrt{9x^4 - 6x^3 + x^2}\right)$$
$$= \frac{1}{2}\left(x^2 + x \pm \sqrt{(3x^2 - x)^2}\right)$$

so $y = 2x^2$ or $y = x - x^2$. Thus $2x^2 = 3$ so $x = \pm\sqrt{3/2}$ or $3 = x - x^2$ i.e. $x^2 - x + 3 = 0$ which has no real roots. Hence $x = \pm\sqrt{3/2}$ are the only real solutions, and the sum of the real solutions is 0.

Answer: 0

Problem 16 Solution
Let $y^2 = \sqrt{x}$, then the expression becomes

$$\left(y^2 + \frac{1}{y^2} - 2\right)^5 = \left(y - \frac{1}{y}\right)^{10},$$

so use the Binomial Theorem to find the coefficient of the constant term, where the constant term is $y^5 \cdot \left(\frac{1}{y}\right)^5 = 1$. Hence the constant term is

$$(-1)^5 \binom{10}{5} = -252.$$

Answer: -252

Problem 17 Solution

Assume that $n^2 + 2018n = k^2$. Completing the square, $(n + 1009)^2 = k^2 + 1009^2$, so $1009^2 = (n + 1009 + k)(n + 1009 - k)$. The maximal n occurs when $n + 1009 + k = 1009^2$ and $n + 1009 - k = 1$. Solve and get $n = 508032$.

Answer: 508032

Problem 18 Solution

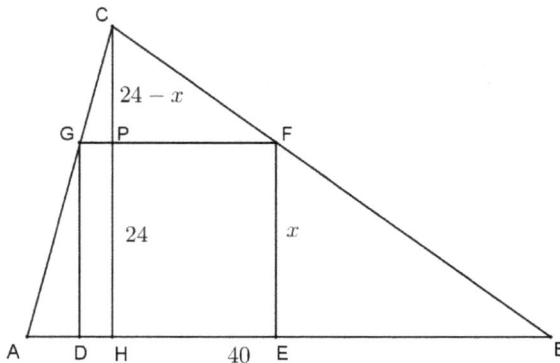

As shown in diagram, let x be the side length of the square, and let P be the intersection of $\overline{CH}$ and $\overline{GF}$. Then $CP = 24 - x$. Since $\overline{GF} \parallel \overline{AB}$, $\triangle CGF \sim \triangle CAB$, thus the altitudes of these two triangles are proportional to their sides, so $\dfrac{24-x}{24} = \dfrac{x}{40}$. Solve for x, we get $x = 15$.

Answer: 15

Problem 19 Solution

Note

$$x^7 + x^6 + x^5 + x^4 + x^3 + x^2 + x + 1 = \frac{x^8 - 1}{x - 1}$$

so after factoring we need to solve

$$(x+1)(x^2+1)(x^4+1) = 0.$$

Hence there is only one real root $x = -1$. Consider the two complex solutions to $x^2 = -1$. Both square to -1 so these contribute a sum of -2. Then note there are 4 complex solutions to $x^4 = -1$, two from $x^2 = i$ and two from $x^2 = -i$. Hence the sum of the squares for these is

$$i + i - i - i = 0$$

so our final answer is -2.

Answer: -2

Problem 20 Solution

By Fermat's Little Theorem, $29^p + 1 \equiv 29 + 1 \equiv 30 \pmod{p}$. But $p \mid 29^p + 1$ implies that $29^p + 1 \equiv 0 \pmod{p}$, hence $30 \equiv 0 \pmod{p}$. It follows that $p \mid 30$, so $p = 2, 3, 5$.

Answer: 10

2.5 ZIML February 2017 Varsity

Below are the solutions from the Varsity ZIML Competition held
in February 2017.
The problems from the contest are available on p.39.

Problem 1 Solution
Let $y = 8$ so we can rewrite the above as

$$x^4 + 8x^3 - 8x - 64 = x^4 + (x^3 - x)y - y^2 = (x^3 - y)(x + y).$$

Hence

$$(x^3 - 8)(x + 8) = 0 \Rightarrow x = 2, -8$$

are the real solutions. The product is $-8 \cdot 2 = -16$.

Answer: -16

Problem 2 Solution
There are 3! ways to arrange the red balls. We know that 2 green
balls must be placed in the gaps between the red balls. This leaves
$10 - 4 = 6$ green balls that can be placed in any of 4 spaces. Using
stars and bars there are

$$\binom{9}{6} = 84$$

ways to arrange the green balls. Hence there are $6 \cdot 84 = 504$
ways in total to arrange the balls.

Answer: 504

Problem 3 Solution
Note that $8 \cdot 252 \equiv 2016 \equiv -1 \pmod{2017}$. Hence we can multiply the equation by -252 to get

$$-252 \cdot 8x \equiv x \equiv -55 \cdot 252 \pmod{2017}.$$

Therefore

$$x \equiv -55 \cdot 252 \equiv -13860 \equiv 259 \pmod{2017}.$$

Therefore the largest solution less than 5000 is $259 + 2015 \cdot 2 = 4293$.

Answer: 4293

Problem 4 Solution

Let O denote the center of the circle and let F, G, and H be respectively the midpoints of $\overline{AB}$, $\overline{CD}$, and $\overline{AC}$. Then as the perpendicular bisectors of any chords go through the center, $\triangle EFO \cong \triangle EGO$. Further, triangles EFO, EGO, EHC, and EHA are all 30-60-90 triangles, as shown in the diagram below.

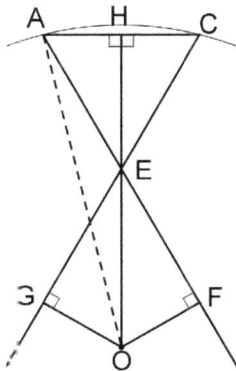

Since $EF = 18$ and $AE = 9$, we have

$$EO = 12\sqrt{3}, EH = 9\sqrt{3} \Rightarrow EH = 21\sqrt{3}.$$

Using the Pythagorean theorem we have the radius AO satisfies

$$AO^2 = 9^2 + \left(21\sqrt{3}\right)^2 = 1404.$$

Hence the area is 1404π so $K = 1404$.

Answer: 1404

Problem 5 Solution
Let $x = 1$:

$$2^7 = a_7 + a_6 + a_5 + a_4 + a_3 + a_2 + a_1 + a_0;$$

let $x = -1$:

$$(-4)^7 = -a_7 + a_6 - a_5 + a_4 - a_3 + a_2 - a_1 + a_0;$$

Adding,
$$-16256 = 2(a_0 + a_2 + a_4 + a_6),$$

so

$$a_0 + a_2 + a_4 + a_6 = -8128.$$

Answer: -8128

Problem 6 Solution
Note that the centers of the four spheres form a regular tetrahedron with side length 20 inches. Hence the height of this tetrahedron is

$$\frac{1}{3} \cdot \sqrt{6} \cdot 20 = \frac{20}{3}\sqrt{6}.$$

Hence the total height is

$$10 + \frac{20}{3}\sqrt{6} + 10 = 20 + \frac{20}{3}\sqrt{6}.$$

Note $2.4 < \sqrt{6} < 2.5$, so

$$16 < \frac{20}{3}\sqrt{6} < 16.\overline{6}.$$

Hence $\lfloor H \rfloor = 36$.

Answer: 36

Problem 7 Solution

We need $16 - x^2 \geq 0$, so $-4 \leq x \leq 4$. In addition, $x^2 + 4x \geq 0$, so $x \leq -4$ or $x \geq 0$. Thus $0 \leq x \leq 4$ or $x = -4$.

It is easy to verify that $x = -4$ is a solution, so now consider the case $0 \leq x \leq 4$:

$$\sqrt{x^2 + 4x} + \sqrt{16 - x^2} \leq 4,$$

squaring both sides (since both sides are nonnegative, it is safe to do so),

$$x^2 + 4x + 16 - x^2 + 2\sqrt{x^2 + 4x} \cdot \sqrt{16 - x^2} \leq 16,$$

which simplified to

$$4x + 2\sqrt{x^2 + 4x} \cdot \sqrt{16 - x^2} \leq 0.$$

Since both terms on the left hand side are nonnegative, they have to be both 0, therefore

$$x = 0.$$

Thus, only $x = 0$ and $x = -4$ are solutions.

Answer: -4

Problem 8 Solution

Call the points the stick is broken X and Y, with $X < Y$. We want the middle piece, which has length $Y - X$ to have length > 2. Hence we are picking a point (X, Y) in the full triangle below, and the area we want is the smaller shaded triangle.

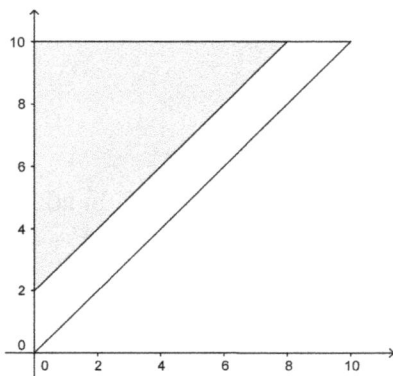

The probability is thus

$$\frac{\frac{1}{2}\cdot 8^2}{\frac{1}{2}\cdot 10^2} = \frac{32}{50} = 64\%.$$

Answer: 64

Problem 9 Solution

The smallest such triangle is shown in the diagram below:

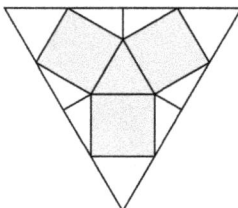

Note the triangle containing the squares is an equilateral triangle. As the triangles outside the shaded squares and triangle are either equilateral triangles or 30-60-90 triangles, the entire triangle has side length

$$2 + \sqrt{3} + \sqrt{3} + 2 = 2\sqrt{3} + 4.$$

Hence the area is

$$(2\sqrt{3} + 4)^2 \cdot \frac{\sqrt{3}}{4} = 12 + 7\sqrt{3}$$

so $R + S + T = 12 + 7 + 3 = 22$.

Answer: 22

Problem 10 Solution

Since the repeating decimal has period 3, we know that

$$0.357357357\ldots = \frac{357}{999} = \frac{119}{333},$$

so $x = 119$.

Answer: 119

Problem 11 Solution

First note that 1 hen and 3 chicks can be bought for $4. Hence one possibility is 0 roosters, 25 hens, and 75 chicks for $100. Further observe that 4 roosters and 3 chicks (a total of 7 chicken) costs the same as 7 hens (again a total of 7 chicken). Hence a second possibility is $0 + 4 = 4$ roosters, $25 - 7 = 18$ hens, and $75 + 3 = 78$ chicks. Doing this twice more, we see that Old McDonald can buy 12 roosters, 4 hens, and 84 chicks for $100 dollars. As we have less than 7 hens remaining, this is the maximum number of roosters Old McDonald can buy.

Answer: 12

Problem 12 Solution

First of all, $x - 4 > 0$, so $x > 4$.

Second, we need

$$1 + \log_{1/3}(x-4) > 0.$$

Thus $\log_{1/3}(x-4) > -1$, so $x-4 < 3$, then $x < 7$. Therefore $4 < x < 7$, which means $L = 3$.

Answer: 3

Problem 13 Solution

Let A denote the event that Keith takes the subway and B denote the event that Keith arrives home between 5:45-5:49 PM. We want to calculate $P(A|B)$. We have

$$P(A|B) = \frac{P(A \cap B)}{P(B)} = \frac{0.50 \cdot 0.45}{0.50 \cdot 0.45 + 0.50 \cdot 0.20} = \frac{0.45}{0.65} = \frac{9}{13}.$$

Hence $M - N = 13 - 9 = 4$.

Answer: 4

Problem 14 Solution

Using the binomial theorem, we get that N is equal to the coefficient of x^{15} in $(1+x)^{15} \cdot (1+x)^{15}$ (collect terms with $x^m \cdot x^k = x^{15}$). Again using the binomial theorem, we must have that N is the coefficient of x^{15} in $(1+x)^{30}$ which is

$$\binom{30}{15} = \frac{30!}{15! \cdot 15!} = \frac{30 \cdot 29 \cdot 28 \cdots 16}{15!}.$$

From here it is clear that the prime 29 is the largest that divides N.

Answer: 29

Problem 15 Solution

$x = \dfrac{1}{3 - \sqrt{7}} = \dfrac{3 + \sqrt{7}}{2}$, and $2 < \sqrt{7} < 3$, so $2 < x < 3$, thus

$$\lfloor x \rfloor = 2, \quad \{x\} = x - 2 = \frac{\sqrt{7} - 1}{2}.$$

Therefore

$$\lfloor x \rfloor + (1+\sqrt{7})\{x\} = 2 + (1+\sqrt{7}) \cdot \frac{\sqrt{7}-1}{2} = 2+3 = 5.$$

Answer: 5

Problem 16 Solution

Rotate $\triangle BCP$ 90 degrees about C so that B ends at point A. Call P' the image of P after the rotation. Note $P'C = 2\sqrt{2}$ and $\angle PCP'$ so $\triangle PCP'$ is an isosceles right triangle. Therefore PP' has length 4. Thus $\triangle APP'$ must be a right triangle and $\angle APP'$ is a right angle. Hence $\angle APC = 135°$. Therefore,

$$AC^2 = 3^2 + (2\sqrt{2})^2 - 2 \cdot 3 \cdot 2\cos(135°) = 29.$$

Finally the area of $\triangle ABC$ is $AC^2/2 = 29/2$ so $N+M = 29+2 = 31$.

Answer: 31

Problem 17 Solution

Assume $x \neq 0$ and set $y = kx$. Solving for x (remember $x \neq 0$ so we can cancel x's) gives

$$x = \frac{18k}{1+k^3} = \frac{20}{1+k^2}.$$

Hence
$$2k^3 - 18k + 20 = 0 \Rightarrow k = 2, -1 \pm \sqrt{6}.$$

Since we want x integral, $k = 2$ and hence $x = 4$. Solving we get $y = 8$ so $(4,8)$ is our solution. The answer is thus $p+q = 4+8 = 12$.

Answer: 12

Problem 18 Solution

By Fermat's little theorem, $3^{16} \equiv 1 \pmod{17}$. Hence we need to calculate $19^{16} \pmod{16}$. Note

$$19^4 \equiv 3^4 \equiv 81 \equiv 1 \pmod{16}.$$

Hence $19^{16} = 16k + 1$ for an integer k. Thus,

$$13^{19^{16}} \equiv 13^{16^k} \cdot 13^1 \equiv 13 \pmod{17}$$

and hence the remainder is 13.

Answer: 13

Problem 19 Solution

Let $A, B, \ldots, G$ be the events that box $1, 2, 3, \ldots 7$ (respectively) get at least 4 balls. We want $n(A \cup B \cup \cdots \cup G)$. Note

$$n(A) = \cdots = n(G) = \binom{6+7-1}{6}$$

using stars and bars (put 4 balls in the respective box, and then arrange the remaining $10 - 4 = 6$ balls in any of the boxes). Similarly we have

$$n(A \cap B) = \cdots = n(F \cap G) = \binom{2+7-1}{2}.$$

Since the intersection of 3 or more of these sets is empty,

$$n(A \cup B \cup C \cup D) = \binom{7}{1} \cdot \binom{12}{6} - \binom{7}{2} \cdot \binom{8}{2} = 5880,$$

as our final answer.

Answer: 5880

Problem 20 Solution

First note $f(t) = t^3 + \sin t$ is an odd function and strictly increasing in $\left[-\frac{\pi}{2}, \frac{\pi}{2}\right]$. We are given that $f(x) = 16$ and

$$f(2y) = 8y^3 + \sin(2y) = 8y^3 + 2\sin y \cos y =$$
$$= -16 = -f(x) = f(-x).$$

As $f(t)$ is strictly increasing, we must have $2y = -x$ and therefore $x + 2y = 0$. Hence $(x + 2y)^3 - \cos(x + 2y) = -1$.

Answer: -1

2.6 ZIML March 2017 Varsity

Below are the solutions from the Varsity ZIML Competition held in March 2017.

The problems from the contest are available on p.45.

Problem 1 Solution

Since $\overline{AB}\|\overline{CD}$, $\triangle ABE \sim \triangle CDE$, with ratio of corresponding sides $10 : 15 = 2 : 3$. Hence $DE : EB = 3 : 2$ and since $\triangle AED$, $\triangle AEB$ share the same height from A, $[AED] : [AEB] = 3 : 2$ so $[AEB] = 24 \cdot \frac{2}{3} = 16$. We can use a similar argument to get $[DEC] = 36$ and $[BEC] = 24$. Hence the total area is 100.

Answer: 100

Problem 2 Solution

Note that

$$31 \equiv -1 \quad (\mathrm{mod}\ 32), \quad 65 \equiv 1 \quad (\mathrm{mod}\ 32),$$

so

$$31^{999} + 65^{100} \equiv (-1)^{999} + 1^{100} \equiv -1 + 1 \equiv 0 \quad (\mathrm{mod}\ 32),$$

so the remainder is 0.

Answer: 0

Problem 3 Solution

Note that we can write 11 as $2 \times 5 + 1$ and also $5 + 6$. Thus, when expanding the multinomial, there will be two terms with x^{11}, namely $\left(x^5\right)^2 x^1$ and $\left(x^5\right)^1 x^6$. We can use the multinomial theorem to find each of their coefficients

$$\frac{10!}{2!1!7!}\left(x^5\right)^2 \cdot x^1 \cdot 2^7 \quad \text{and} \quad \frac{10!}{1!6!3!}\left(x^5\right)^1 \cdot x^6 \cdot 2^3.$$

Hence, the coefficient of x^{11} is

$$\frac{10!}{2!1!7!} \cdot 2^7 + \frac{10!}{1!6!3!} \cdot 2^3 = 52800.$$

Answer: 52800

Problem 4 Solution

By Vieta's formulas, $x_1 + x_2 = 2, x_1 x_2 = \dfrac{k}{4}$. Thus

$$\frac{1}{x_1} + \frac{1}{x_2} = \frac{x_1 + x_2}{x_1 x_2} = \frac{8}{k} = \frac{8}{3}.$$

Therefore $k = 3$.

Answer: 3

Problem 5 Solution

Expanding the right side we get $|x^2 + 6x + 5|$, so with the substitution $y = x^2 + 6x + 1$ we get $|y| = |y + 4|$. Hence we must have $y = -y - 4$ so $y = -2$. Thus we need $x^2 + 6x + 1 = -2$ or $x^2 + 6x + 3 = 0$. The quadratic formula gives us $-3 \pm \sqrt{6}$ as roots. Hence, $|A| = 3$ and $|B| = 6$, so $|A| + |B| = 9$.

Answer: 9

Problem 6 Solution

$60 = 2^2 \cdot 3 \cdot 5$, so numbers with exactly 60 factors will be of the form

$$p^4 \cdot q^2 \cdot r \cdot s \text{ or } p^4 \cdot q^3 \cdot r^2 \text{ or } p^5 \cdot q^4 \cdot r \text{ etc.}$$

for primes p, q, r, s. Since we want the smallest integer, $p = 2$, $q = 3$, $r = 5$, $s = 7$ and we see the smallest such integer is

$$2^4 \cdot 3^2 \cdot 5 \cdot 7 = 5040.$$

Answer: 5040

Problem 7 Solution

Using the Law of Cosines we have

$$BC^2 = 9 + 8 - 12\sqrt{2}\cos(45°) = 5$$

so $BC = \sqrt{5}$. Similarly, using the Law of Cosines again we can calculate $\cos(B) = \dfrac{\sqrt{5}}{5}$. Since $\sin^2(B) + \cos^2(B) = 1$ we get $\sin(B) = \dfrac{2\sqrt{5}}{5}$. Hence

$$\tan(B) = \frac{\sin(B)}{\cos(B)} = 2.$$

Answer: 2

Problem 8 Solution

First arrange the 6 red cards, there are 6! ways to do it. Now place 2 black cards in between each of the red cards (5 spaces). Now use stars and bars to place the remaining $15 - 2 \times 5 = 5$ black cards (in 7 spaces between and on the ends of the red cards). Hence we have

$$6! \cdot \binom{5 + 7 - 1}{5} = 332640$$

different ways of arranging the cards.

Answer: 332640

Problem 9 Solution

As shown in the diagram below, the two dotted line segments have the same length and are perpendicular to each other, so the big triangle is a 45-45-90 triangle.

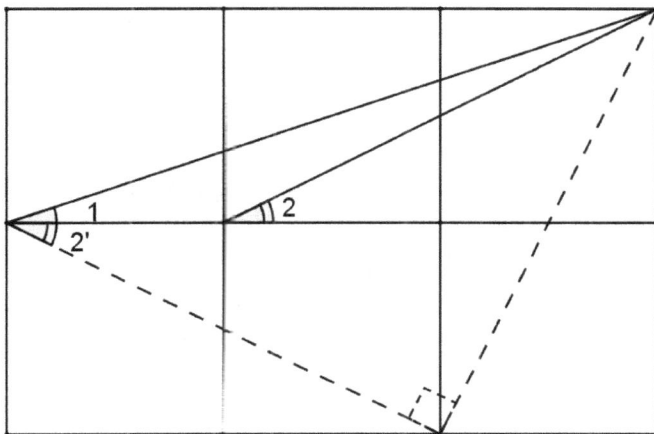

Hence $\angle 1 + \angle 2 = \angle 1 + \angle 2' = 45°$.

Answer: 45

Problem 10 Solution

The number 1 can only be on the numerator, and 2 can only be on the denominator. Everything else can be either on top or on the bottom: $2^6 = 64$. However, since $3 \times 8 = 4 \times 6$, the result is repeated when 3×8 is on top and 4×6 is on the bottom, and vice versa. In each of those cases there are $2^2 = 4$ choices to place 5 and 7, so these 4 choices should be subtracted from 64, getting 60 as the final answer.

Answer: 60

Problem 11 Solution

Note first that the domain of $\log(x-20) + \log(30-x) = \log[(x-20)(30-x)]$ is $20 < x < 30$. Trying $x = 21$ we have $\log(9) < 1$ but if $x = 22$ we have $\log(16) > 1$. By symmetry, we have $x = 28$ does not work, but $x = 29$ does work. Hence there are two integer solutions.

Answer: 2

Problem 12 Solution
We have $60^4 = 2^8 \cdot 3^4 \cdot 5^4$. Perfect square divisors will be of the form $2^{2l}3^{2m}5^{2n}$ for $l = 0, 1, 2, 3, 4; m, n = 0, 1, 2$. Thus there are $5 \cdot 3 \cdot 3 = 45$ perfect square divisors. These come in pairs, each multiplying out to 60^4 except for 60^2 (which is paired with itself). Hence the product of all the perfect square divisors is $(60^4)^{22} \cdot (60^2) = 60^{4 \cdot 22 + 2} = 60^{90}$. Therefore $K = 90$.

Answer: 90

Problem 13 Solution
Let A be the event that the rolls add up to 6. Then $P(T) = P(H) = 1/2$, $P(A|T) = 5/36$, and $P(A|H) = 1/6$. Using Bayes' Theorem we have that

$$P(T|A) = \frac{\dfrac{1}{2} \cdot \dfrac{5}{36}}{\dfrac{1}{2} \cdot \dfrac{5}{36} + \dfrac{1}{2} \cdot \dfrac{1}{6}} = \frac{5}{11}$$

Hence $Q - P = 6$.

Answer: 6

Problem 14 Solution
Use $[XYZ]$ to denote the area of $\triangle XYZ$.
Since $[CBD] = [BAE] = [ACF]$, we get

$$\frac{[CBD]}{[ABC]} = \frac{[BAE]}{[ABC]} = \frac{[ACF]}{[ABC]},$$

thus

$$\frac{212}{CA} = \frac{208}{BC} = \frac{204}{AB}.$$

Simplifying,

$$\frac{53}{CA} = \frac{52}{BC} = \frac{51}{AB}.$$

Since AB, BC, CA are all integers, the minimum possible values are $AB = 51$, $BC = 52$, $CA = 53$, so by Heron's formula,

$$[ABC] = \sqrt{78(78-51)(78-52)(78-53)} = 1170.$$

Answer: 1170

Problem 15 Solution

Since $\lfloor -1.77 \rfloor = -2$, the equation can be simplified to

$$\lfloor -1.77x \rfloor = -2x.$$

Hence as $-2x$ must be an integer, we see x is a multiple of $1/2$. Also $\{-1.77x\} = -1.77x - \lfloor -1.77x \rfloor$, which means

$$\lfloor -1.77x \rfloor = -1.77x - \{-1.77x\},$$

so $-1.77x - \{-1.77x\} = -2x$, then $0.23x = \{-1.77x\}$. From $0 \le \{-1.77x\} < 1$, we get $0 \le 0.23x < 1$, so

$$0 \le x < \frac{100}{23}.$$

Thus the possible values for x are $0, 0.5, 1, 1.5, 2, 2.5, 3, 3.5, 4$. Hence the sum of these values is 18.

Answer: 18

Problem 16 Solution

The equation $x^2 + x + 1 = 0$ implies that $x^3 = 1$, so x is the cube root of unity. Also, divide both sides by x, we get $x + 1 + \frac{1}{x} = 0$, so $x + \frac{1}{x} = -1$. Furthermore, $x^2 + \frac{1}{x^2} = x^2 + x = -1$, and $x^3 + \frac{1}{x^3} = 2$.

In general, $x^k + \dfrac{1}{x^k} = 2$ if k is a multiple of 3, and -1 otherwise. Therefore,

$$\left(x + \frac{1}{x}\right)^2 + \left(x^2 + \frac{1}{x^2}\right)^2 + \cdots + \left(x^{27} + \frac{1}{x^{27}}\right)^2 =$$
$$= 1 + 1 + 4 + 1 + 1 + 4 + \cdots + 1 + 1 + 4 = 54.$$

Answer: 54

Problem 17 Solution
Note that

$$\overline{abcd} + \overline{bcde} \equiv 1000a + 1100b + 110c + 11d + e \pmod{11}$$
$$\equiv e - a \pmod{11}.$$

Hence the sum is divisible by 11 if $e - a$ is divisible by 11. Since a, e are between $0 - 9$, this means $a = e$. Since $a \neq 0$, there are 9 choices for a, and e has to equal a. The digits b, c, d can be any of $0, 1, \ldots, 9$ (10 choices each) for a total of $9 \cdot 10 \cdot 10 \cdot 10 \cdot 1 = 9000$ numbers.

Answer: 9000

Problem 18 Solution
First note that since AD is parallel to BC, we have $\angle DAE = \angle BFA$, call this angle θ. We have $\dfrac{AD}{AE} = \dfrac{2}{AE} = \cos(\theta)$ and $\dfrac{AB}{AF} = \dfrac{2}{AF} = \sin(\theta)$. Hence

$$\frac{1}{AE^2} + \frac{1}{AF^2} = \frac{\cos^2(\theta) + \sin^2(\theta)}{2^2} = \frac{1}{4},$$

as needed.

Answer: 0.25

Problem 19 Solution

Let A and B be the events where 1 is adjacent to 2 and 0, respectively. Then $n(A) = 2! \cdot 4 \cdot 4!$ if we think of $1, 2$ as a pair, as there are 4 choices for were the 0 goes. For $n(B)$ consider the groupings 10 and 01 separately to get $n(B) = 5! + 4 \cdot 4!$ (for 01, the pair 01 cannot be first, for 10 the digit 0 is automatically not first). For $A \cap B$, group $2, 1, 0$ with 1 in the middle, again considering 210 and 012 separately, so $n(A \cap B) = 4! + 3 \cdot 3!$. Using PIE, the total number of ways to arrange the digits will be given by

$$2! \cdot 4 \cdot 4! + 5! + 4 \cdot 4! - (4! + 3 \cdot 3!) = 366$$

Answer: 366

Problem 20 Solution

Using Cauchy-Schwarz Inequality:

$$
\begin{aligned}
& f(a,b,c) + 10 \\
= & \frac{a}{b+c} + \frac{4b}{c+a} + \frac{5c}{a+b} + 10 \\
= & (a+b+c) \left(\frac{1}{b-c} + \frac{4}{c+a} + \frac{5}{a+b} \right) \\
= & \frac{1}{2}((b+c) + (c-a) + (a+b)) \left(\frac{1}{b+c} + \frac{4}{c+a} + \frac{5}{a+b} \right) \\
\geq & \frac{1}{2}(1 + 2 + \sqrt{5})^2 \\
= & 7 + 3\sqrt{5}.
\end{aligned}
$$

Hence $f(a,b,c) \geq 3\sqrt{5} - 3$.

Since $3\sqrt{5} = \sqrt{45}$, we get

$$6 < 3\sqrt{5} < 7,$$

thus
$$3 < 3\sqrt{5} - 3 < 4,$$

so $L = 4$.

Answer: 4

2.7   ZIML April 2017 Varsity

Below are the solutions from the Varsity ZIML Competition held in April 2017.
The problems from the contest are available on p.51.

Problem 1 Solution

Let r be the radius of $\odot G$. Let A, B, D be the centers of the smaller circles, as shown in the diagram.

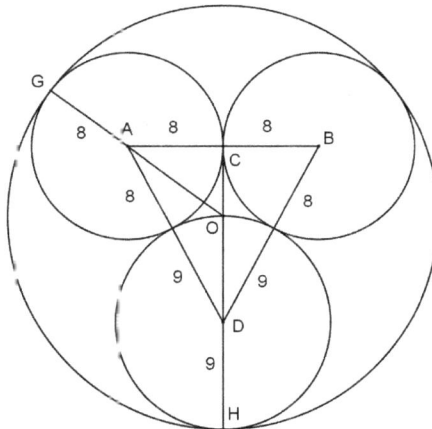

So we have $OA = r - 8$ and $OD = r - 9$. Also, $\triangle ADC$ is a right triangle, and then $DC = \sqrt{17^2 - 8^2} = 15$. It follows that $OC = DC - OD = 24 - r$.

We also know that $\triangle OAC$ is a right triangle, therefore $AC^2 + OC^2 = OA^2$, that is,

$$8^2 + (24 - r)^2 = (r - 8)^2,$$

which simplifies to

$$576 = 32r,$$

hence $r = 18$.

Answer: 18

Problem 2 Solution

$2836 \equiv 4582 \equiv 5164 \equiv 6522 \pmod{m}$, so $1746 \equiv 582 \equiv 1358 \equiv 0 \pmod{m}$. This means m is a common factor of 1746, 582, and 1358. Since $\gcd(1746, 582, 1358) = 194$, $m \mid 194$. So m can be 2, 97, or 194. If $m = 2$, the remainder $r = 0$ which is not a solution. If $m = 97$, $r = 23$. if $m = 194$, $r = 120$. Hence $r = 23$ or 120 so the sum is 143.

Answer: 143

Problem 3 Solution

It is easiest to use complementary counting. Using stars and bars, there are

$$\binom{21+4-1}{21} = \binom{24}{21} = 2024$$

total ways to distribute the tickets. Note that one group gets more tickets than the other three groups combined if and only if one group gets at least 11 tickets. In this case, there are 4 choices for which group gets 11 tickets, and then we distribute the remaining 10 tickets using star and bars (among all four groups). There are thus

$$4\binom{10+4-1}{10} = 1144$$

was for one group to get more tickets than the other three combined. This gives a final answer of

$$2024 - 1144 = 880.$$

Answer: 880

Problem 4 Solution

Using the logarithm rules we can rewrite the equation as

$$(\log_{10}(x))^2 = \log_{10}(x) + 2.$$

Setting $z = \log_{10}(x)$, we have the quadratic $z^2 - z - 2 = 0$ whose roots are $z = 2, z = -1$. Hence the solutions are $x = 100$ and $x = 1/10$, with product 10.

Answer: 10

Problem 5 Solution
We first have $[ABD] = [DEA]$ (area of $\triangle ABD$ and $\triangle DEA$) as they have the same height and base. Hence, after removing the shared $\triangle AFD$ we see the two shaded triangles have the same area. Hence it is enough to find $[DEF]$. Since E is a midpoint, $[DEC] = [BED] = [ABCD] \div 4$. Since $ABCD$ is a parallelogram and $BE = AD/2$, $\triangle AFD \sim \triangle EFB$, with ratio of sides $2 : 1$. Using this information, we have $[DFE] = 2[BEF]$. Therefore, $[DFE] = [ABF] = [ABCD] \div 6 = 72 \div 6 = 12$. Hence the two shaded regions have a combined area of $12 \times 2 = 24$.

Answer: 24

Problem 6 Solution
Note mod 3 are numbers are

$$1, 2, 0, 1, 2, 0, 0, 0, 0.$$

Further, for two numbers to be divisible by 3, one of them must be 0 (mod 3). Hence we need to arrange 5 (distinct) numbers that are $\equiv 0$ (mod 3) and 4 (distinct) numbers that are $\not\equiv 0$ (mod 3) such that the numbers not 0 (mod 3) are separated. There are

$$5! \cdot 6 \cdot 5 \cdot 4 \cdot 3 = 43200$$

ways to do this, as after we arrange the 5 numbers there are 6 spots for the remaining 4.

Answer: 43200

Problem 7 Solution

We have

$$\begin{aligned}
&\gcd(20162017, 20152016) \\
=\ &\gcd(20152016, 20162017 - 20152016) \\
=\ &\gcd(20152016, 10001).
\end{aligned}$$

Then note $20152016 = 10001 \times 2015 + 1$ so the above greatest common divisor is equal to $\gcd(10001, 1) = 1$.

Answer: 1

Problem 8 Solution

a, b, c satisfy $x^3 + 4x^2 - 3x - 3 = 0$ or $x^3 = -4x^2 + 3x + 3$, so

$$a^3 + b^3 + c^3 = -4(a^2 + b^2 + c^2) + 3(a + b + c) + 3 \cdot 3.$$

By Vieta's Formulas, $a + b + c = -4$, and

$$a^2 + b^2 + c^2 = (a + b + c)^2 - 2(ab + ac + bc) = 16 - 2(-3) = 22.$$

Hence $a^3 + b^3 + c^3 = -4(22) + 3(-4) + 9 = -91$.

Answer: -91

Problem 9 Solution

Suppose the side length of the triangle is 1. Then the area of the equilateral triangle equals $\dfrac{\sqrt{3}}{4}$. Also we know that the area of a triangle equals rs, where r is a inradius and s is the semiperimeter. In this case, $s = \dfrac{3}{2}$, so

$$r = \frac{\sqrt{3}/4}{3/2} = \frac{\sqrt{3}}{6}.$$

Let θ be the angular measure of the arc, we have

$$1 = \frac{\sqrt{3}}{6}\theta,$$

therefore $\theta = 2\sqrt{3}$. Hence $A + B + C = 2 + 3 + 1 = 6$.

Answer: 6

Problem 10 Solution

There are $\binom{30}{2}$ ways to pick 2 of the marbles. Factoring, we have $n^2 - 20n + 103 = (n-10)^2 + 3$, so by symmetry, there are 9 pairs of labels, $(9, 11), (8, 12), \ldots, (1, 19)$, that correspond to marbles that weigh the same. Hence the probability is

$$9 \div \binom{30}{2} = \frac{18}{30 \cdot 29} = \frac{3}{145},$$

so $P + Q = 3 + 145 = 148$.

Answer: 148

Problem 11 Solution

On the first pass, Carrie erases all the odd numbers, leaving only the even numbers (multiples of 2): $2, 4, 6, 8, \ldots, 150$. After the second pass, only the multiples of 4 remain: $4, 8, 12, \ldots, 148$. In general, after the nth pass, only the multiples of 2^n remain. Hence, after the 7th pass, only 128 will remain.

Answer: 128

Problem 12 Solution

Squaring, we get $x^2 - (14 + 2a)x + a^2 + 6a + 25 = 0$. The discriminant $\Delta = 32(a+3) \geq 0$, so $a \geq -3$. By the quadratic formula, $x = 7 + a \pm 2\sqrt{6 + 2a}$. Thus $6 + 2a$ must be a square (and also even). The even squares are $0, 4, 16, 36, \ldots$, and the corresponding values for a is $-3, -1, 5, 15, \ldots$. Hence $a = 5, -1,$ or 3. Checking $a = 5$, $\sqrt{2x-4} - \sqrt{x+5} = 1$ has exactly one integer root ($x = 20$), so our answer is 5.

Answer: 5

Problem 13 Solution

Note by symmetry the above is equal to

$$\binom{5}{5} + \binom{6}{5} + \binom{7}{5} + \binom{8}{5} + \cdots + \binom{15}{5}$$

which equals $\binom{16}{6}$ using the hockey-stick identity. Thus our answer is

$$\binom{16}{6} = \frac{16 \cdot 15 \cdot 14 \cdot\cdot 13 \cdot 12 \cdot 11}{6!} = 4 \cdot 14 \cdot 13 \cdot 11 = 8008.$$

Answer: 8008

Problem 14 Solution

First note using Fermat's Little Theorem, $3^{10} \equiv 1 \pmod{11}$. Further, $3^4 \equiv 1 \pmod{10}$. Therefore,

$$3^3 + 3^{3^2} + 3^{3^3} + 3^{3^4} \equiv 3^{3^5} + \cdots + 3^{3^8} \pmod{11}$$

Hence

$$3^3 + 3^{3^2} + 3^{3^3} + \cdots + 3^{3^8} \equiv 2 \cdot (3^3 + 3^{3^2} + 3^{3^3} + 3^{3^4}) \pmod{11}.$$

Using patterns we have $3^3 \equiv 5 \pmod{11}$, $3^{3^2} \equiv 3^9 \equiv 4 \pmod{11}$, $3^{3^3} \equiv 3^7 \equiv 9 \pmod{11}$, and $3^{3^4} \equiv 3^1 \equiv 3 \pmod{11}$. Thus, the final answer is $2 \cdot (5 + 4 + 9 + 3) \equiv 9 \pmod{11}$.

Answer: 9

Problem 15 Solution

Let $a = 2\cos\alpha + 2i\sin\alpha$ and $b = 2\cos\beta + 2i\sin\beta$, then $2\cos\alpha + 2i\sin\alpha + 2\cos\beta + 2i\sin\beta + 2 = 0$. Considering real and imaginary parts, we get $\cos\alpha + \cos\beta = -1$ and $\sin\alpha + \sin\beta = 0$. As $\sin\alpha = -\sin\beta$, we either have (1) $\alpha = 2k\pi - \beta$ or (2) $\alpha =$

$(2k+1)\pi+\beta$. In case (1), $\cos\alpha = \cos\beta$, thus they are both $-\dfrac{1}{2}$.

Similarly get $\sin\alpha = \pm\dfrac{\sqrt{3}}{2}$, so $a,b = -1\pm\sqrt{3}$.

In case (2), $\cos\alpha = -\cos\beta$ which is impossible.

Hence $|\text{Im}(a-b)| = 2\sqrt{3} = \sqrt{12}$, so $S = 12$.

Answer: 12

Problem 16 Solution
Suppose the entire volume is 1. Let P and Q be respectively the intersections of $\overline{AF},\overline{CD}$ and $\overline{BF},\overline{CE}$. Using similar triangles, note in fact P and Q are midpoints of all the respective segments. The four parts are bounded by (i) C,F,P,Q, (ii) A,B,C,P,Q, (iii) D,E,F,P,Q, and (iv) A,B,D,E,P,Q. The triangular pyramid $F-ABC$ shares a base with the prism, hence has volume $\dfrac{1}{3}$.
Consider the same pyramid with height from C (so the height is in the plane containing $\triangle BFA$). This height is the same if we look at pyramid $C-PFQ$ (with base $\triangle PFQ$). As $\triangle PFQ \sim \triangle BFA$ with ratio of sides $1:2$, we know that the volume of $C-PFQ$ (region (i)) is
$$\frac{1}{3}\cdot\frac{1}{4} = \frac{1}{12}.$$
Thus region (ii) (and similarly region (iii)) have volume
$$\frac{1}{3} - \frac{1}{12} = \frac{1}{4}.$$
Therefore region (iv) has volume
$$1 - \frac{1}{12} - \frac{1}{4} - \frac{1}{4} = \frac{5}{12}.$$
Hence the volumes of the largest (region (iv)) and the smallest (region (i)) have ratio $5:1$ so our answer $K = 5$.

Answer: 5

Problem 17 Solution

We get that $M = \dfrac{2017 \times 2016}{2} = 2017 \times 1008$, and since 2017 is prime, $\gcd(2017, 1008) = 1$. By Wilson's Theorem,

$$2016! \equiv -1 \equiv 2016 \pmod{2017}.$$

Also we know that

$$2016! \equiv 0 \equiv 2016 \pmod{1008},$$

therefore by Chinese Remainder Theorem,

$$2016! \equiv 2016 \pmod{2017 \times 1008}.$$

Answer: 2016

Problem 18 Solution

Let $y = \sqrt{2}$ to rewrite as a quadratic in y: $y^2 + (1 + x - x^2)y + (x - x^3)$. Noting that $x - x^3 = x(1 - x^2)$ we see the expression factors as $(y + x)(y + 1 - x^2) = 0$. Hence $\sqrt{2} = -x$ so $x = -\sqrt{2}$ or $\sqrt{2} = 1 - x^2$ so $x = \pm\sqrt{1 + \sqrt{2}}$. Therefore $a = 1$, $b = 2$ and our answer is $1 + 2 = 3$.

Answer: 3

Problem 19 Solution

Suppose the second stick has length x and third stick length y. Using the triangle inequality, we need (i) $1 + x > y$ or $y < x + 1$, (ii) $1 + y > x$ or $y > x - 1$, and (iii) $x + y > 1$ or $y > 1 - x$. Graphing this in a 2×3 rectangle, we get that (x, y) must be in the shaded region shown below:

The full triangle has area 6, while the shaded region has area $6 - 2 - \frac{1}{2} - \frac{1}{2} = 3$, so the probability is $\frac{3}{6} = 50\%$, so $L = 50$.

Answer: 50

Problem 20 Solution

Reflect point B across line $\overline{CE}$ to point B'.

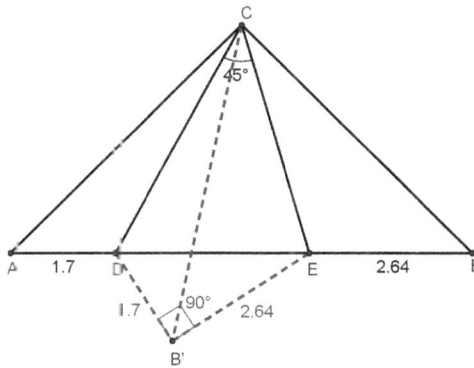

Then $B'E = BE = 2.64$, $CB' = CB = CA$, and $\angle DCB' = \angle DCA$, thus triangles DCB' and DCA are congruent. Therefore $DB' = DA = 1.7$. Also $\angle DB'E = 90°$, so by Pythagorean theorem, $DE = 3.14$.

Answer: 3.14

2.8   ZIML May 2017 Varsity

Below are the solutions from the Varsity ZIML Competition held in May 2017.

The problems from the contest are available on p.57.

Problem 1 Solution
$AB = BC = CD = AD = 20$, let $AF = FE = x$. Then $DF = 20 - x$ and $CF = CE + EF = BC + EF = 20 + x$. In the right triangle CDF,

$$DF^2 + CD^2 = CF^2 \Rightarrow (20 - x)^2 + 20^2 = (20 + x)^2.$$

Solve and get $x = 5$, so $DF = 20 - x = 15$.

Answer: 15

Problem 2 Solution
Let $m = 2121 \cdots 212$ be k pairs of 21 followed by a 2. Then $m \equiv k(2 - 1) + 2 \equiv k + 2 \pmod{11}$. Hence $n \equiv 2 + 3 + 4 + 5 + 6 + 7 \equiv 5 \pmod{11}$.

Answer: 5

Problem 3 Solution
First, pick an odd digit for the last digit (to ensure the number is odd), 5 choices. Second, pick a nonzero digit from the remaining 9 digits as the leading digit (to ensure the number has five digits), 8 choices. Third, pick any 3 digits out of the 8 remaining digits and arrange them on the tens, hundreds, and thousands digits, $8 \times 7 \times 6$ choices. This gives a final answer of

$$5 \times 8 \times (8 \times 7 \times 6) = 13440.$$

Answer: 13440

Problem 4 Solution

Since x_1 and x_2 are integers, the possible values for $x_1^2 + 1$ and $x_2^2 + 1$ are $1, 10$ and $2, 5$. In the first case, $x_1 = 0, x_2 = \pm 3$; and in the second case, $x_1 = \pm 1$ and $x_2 = \pm 2$. The order of x_1 and x_2 does not matter, since we only need to find the possible pairs of (m, n). By Viete's formulas, $-m = x_1 + x_2, 2 - n = x_1 x_2$, so there are 6 different pairs for (m, n).

Answer: 6

Problem 5 Solution

Suppose the altitudes (from A, B, C respectively) are h_A, h_B, h_C with $h_A : h_B : h_C = 10 : 10 : 13$. If a, b, c are the opposite sides, we then have, after canceling $1/2$ from area calculations: $a \cdot h_A = b \cdot h_B = c \cdot h_C$. We therefore have that $b : a = h_A : h_B = 10 : 10$ and $c : b = h_B : h_C = 13 : 10$. Hence, the sides are in ratio $c : b : a = 10 : 13 : 13$. If the perimeter is 72, then we can calculate the sides as $20, 26, 26$. As $(5, 12, 13)$ and thus $(10, 24, 26)$ are Pythagorean triples, we see that the altitude from the side of length 20 is 24. Hence the triangle as area $\dfrac{1}{2} \times 20 \times 24 = 240$.

Answer: 240

Problem 6 Solution

There are 4 possible ways to partition 7 as the sum of 3 odd numbers:

$$11 = 9 + 1 + 1 = 7 + 3 + 1 = 5 + 5 + 1 = 5 + 3 + 3.$$

For the case with $9, 1, 1$ balls in each box, we must choose which box has 9 balls and then arrange the balls into the boxes, giving

$$3 \cdot \binom{11}{9} \binom{2}{1} \binom{1}{1}$$

ways. Similarly for the case of $7,3,1$ balls in each box we choose which box gets $7,3,1$ balls and then arrange the balls to get

$$3! \cdot \binom{11}{7} \binom{4}{3} \binom{1}{1}$$

ways. For the $5,5,1$ case we have

$$3 \cdot \binom{11}{5} \binom{6}{5} \binom{1}{1}$$

ways and finally the $5,3,3$ case gives

$$3 \cdot \binom{11}{5} \binom{6}{3} \binom{3}{3}$$

ways. Hence our final answer is

$$3 \cdot \binom{11}{9} \binom{2}{1} \binom{1}{1} + 3! \cdot \binom{11}{7} \binom{4}{3} \binom{1}{1}$$
$$+ 3 \cdot \binom{11}{5} \binom{6}{5} \binom{1}{1} + 3 \cdot \binom{11}{5} \binom{6}{3} \binom{3}{3}$$
$$= 44286$$

total outcomes.

Answer: 44286

Problem 7 Solution
Let $n^3 = 5p+1$, then $5p = n^3 - 1 = (n-1)(n^2+n+1)$. Either $n-1 = 5$ or $n^2+n+1 = 5$, and the latter doesn't give integer solutions for n. If $n-1 = 5$, $n = 6$, and $p = n^2+n+1 = 43$.

Answer: 43

Problem 8 Solution
Consider the function $f(x) = \log_{100}(\log_{1/10}(\log_{10}(x^2 - 1)))$. Note $\log_{10}(x^2 - 1)$ is defined as long as $|x| \geq 1$. For clarity let $y =$

$\log_{10}(x^2 - 1)$ and $z = \log_{1/10}(\log_{10}(x^2 - 1))$. For $\log_{1/10}(y)$ to be defined we need $y > 0$ so $\log_{10}(x^2 - 1) > 0$ so $x^2 - 1 > 10$ and hence $|x| \geq \sqrt{11}$. Lastly for $\log_{100}(z)$ to be defined we need $z > 0$ so we need $\log_{1/10}(y) > 0$. Hence $y < 1$ so $x^2 - 1 < 1$ and $|x| < \sqrt{2}$. Combining all the restrictions we see $\sqrt{2} < |x| < \sqrt{11}$ is the domain of $f(x)$. As $\sqrt{2} \approx 1.41$ and $\sqrt{11} \approx 3.32$, we have that $n = \pm 2, \pm 3$ are the only integers in the domain. Hence our answer is 4.

Answer: 4

Problem 9 Solution
Let E be the intersection of $\overline{AC}$ and the circle. Extend $\overline{CA}$ to intersect the circle at F. So $AE = AF = AB = 71, CF = 84 + 71 = 155$, and $CE = 84 - 71 = 13$. By Power of a Point, $CB \cdot CD = CF \cdot CE = 155 \times 13 = 2015$. Since the lengths of $\overline{BD}$ and $\overline{DC}$ are integers, so is the length of $\overline{BC}$. Since $\overline{CE}$ goes through A, the center of the circle, $CF > CB$. Since $2015 = 5 \times 13 \times 31$, the only way that 2015 is expressed as the product of two factors, and both factors are less than 155, is $2015 = 65 \times 31$. Thus $BC = 65$ and $CD = 31$. The final answer: 65.

Answer: 65

Problem 10 Solution
Working directly,

$$6, 17, 28, 39, 50, 61, 72, 83, 94, \ldots \equiv 6 \pmod{11}.$$

Of these, 94 is the smallest that is 4 (mod 9). Therefore numbers of the form $94 + 99k$ will be satisfy the restrictions both mod 9 and mod 11. Listing these we have

$$94, 193, 292, \ldots,$$

and we see that $292 \equiv 2 \pmod 5$. Hence all solutions are of the form $292 + 495k$ for integers k, as $\mathrm{lcm}(5, 9, 11) = 495$. Note

$10000 \div 495 = 20\ R\ 100$, so the smallest solution less than 10000 is

$$19 \cdot 495 + 292 = 9697.$$

Answer: 9697

Problem 11 Solution

Let $x = r\cos\theta$, $y = r\sin\theta$, where $0 \le r \le \sqrt{2}$, then

$$\begin{aligned}|x^2 - 2xy - y^2| &= |r^2(\cos^2\theta - 2\cos\theta\sin\theta - \sin^2\theta)| \\ &= r^2|\cos 2\theta - \sin 2\theta| \\ &= \sqrt{2}r^2\cos(2\theta + 45^\circ) \le 2\sqrt{2}.\end{aligned}$$

As $2\sqrt{2} = \sqrt{8}$, we have $M = 8$.

Answer: 8

Problem 12 Solution

Let A be the event that you roll a 5 and B the event that you get exactly 5 heads. First note

$$P(B) = \frac{1}{6} \cdot \left(\frac{1}{2}\right)^5 + \frac{1}{6} \cdot \binom{6}{5}\left(\frac{1}{2}\right)^6$$

as you either roll a 5 and then get 5 heads, or roll a 6 and then get 5 heads and 1 tails (in some order). Similarly,

$$P(A \cap B) = \frac{1}{6} \cdot \left(\frac{1}{2}\right)^5$$

as now we also are given that we rolled a 5. Hence,

$$P(A|B) = \frac{P(A \cap B)}{P(B)} =$$

$$= \left[\frac{1}{6} \cdot \left(\frac{1}{2}\right)^5\right] \div \left[\frac{1}{6} \cdot \left(\frac{1}{2}\right)^5 + \frac{1}{6} \cdot \binom{6}{5}\left(\frac{1}{2}\right)^6\right].$$

Multiplying numerator and denominator by $6 \cdot 2^6$ we get

$$F(A|B) = \frac{2}{2+6} = \frac{1}{4}$$

so $P+Q = 1+4 = 5$.

Answer: 5

Problem 13 Solution

Grouping the outside two terms and the inside two terms: $(x^2 + 8x+7)(x^2+8x+15)+15$. Substituting $y = x^2+8x+7$ we have $y^2+8y+15 = (y+5)(y+3)$. Rewriting in terms of x and further factoring gives $(x+2)(x+6)(x^2+8x+10)$. Hence $x = -2, x = -6$ are the only integer solutions, with sum $(-2)+(-6) = -8$.

Answer: -8

Problem 14 Solution

Let the tetrahedron be $A - BCD$, where $\triangle BCD$ is the equilateral triangle with side length 1. Let G be the center of $\triangle BCD$, so $\overline{AG} \perp$ plane BCD. Since $BG = \sqrt{3}/3$ and $AB = 2$,

$$AG = \sqrt{2^2 - \frac{1}{3}} = \frac{\sqrt{33}}{3}.$$

Therefore the volume is

$$\frac{1}{3} \cdot \frac{\sqrt{3}}{4} \cdot \frac{\sqrt{33}}{3} = \frac{\sqrt{11}}{12},$$

so $P+Q = 11+12 = 23$.

Answer: 23

Problem 15 Solution

There are $4^3 = 64$ total ways to complete the test. Hence there are $64 \cdot 63$ total ways for David and John to complete the test differently. Similarly there are $2 \cdot 1 \cdot 63$ ways for one of them to get a perfect score, as then the other must have one of the 63 wrong sets of answer choices. Thus the probability is

$$\frac{2 \cdot 63}{64 \cdot 63} = \frac{1}{32}$$

so $M - N = 32 - 1 = 31$.

Answer: 31

Problem 16 Solution

We have that $140 = 2^2 \cdot 5 \cdot 7$. Therefore x is either 2, 5, or 7.

First assume $x = 7$, so we have $(y+7)(y+z) = 20$. As $y, z \geq 2$, we have $(y+7)(y+z) \geq 9 \cdot 4 = 36$ so this is impossible.

Next assume $x = 5$, so we have $(y+5)(y+z) = 28$. As $y, z \geq 2$, the first term must be at least 7 and the second term must be at least 4, leaving only $28 = 7 \cdot 4$ which implies $y = 2$ and $z = 2$.

Lastly assume $x = 2$, so we have $(y+2)(y+z) = 70$. As $y, z \geq 2$, we look at the factor pairs: $(14,5)$, $(10,7)$, $(7,10)$, and $(5,14)$. Of these, only $(7,10)$ and $(5,14)$ work, leading to pairs $(5,5)$ and $(3,11)$.

Hence there are 3 triples in total: $(5,2,2)$, $(2,5,5)$, and $(2,3,11)$.

Answer: 3

Problem 17 Solution

Note that

$$k \cdot \binom{10}{k} = 10 \cdot \binom{9}{k-1} \text{ for } k = 1, 2, 3, \ldots, 10.$$

Hence

$$\frac{1}{1024}\left(1\cdot\binom{10}{1}+2\cdot\binom{10}{2}+3\cdot\binom{10}{3}+\cdots+10\cdot\binom{10}{10}\right)$$
$$=\frac{10}{1024}\left(\binom{9}{0}+\binom{9}{1}+\binom{9}{2}+\cdots+\binom{9}{9}\right)$$
$$=\frac{10}{1024}\cdot2^9=5$$

Answer: 5

Problem 18 Solution

Rotate $\triangle APC$ $60°$ about A so that C reaches B and P moves to P' as in the diagram below.

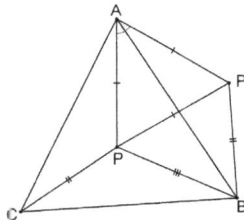

Then we have that $\triangle BPP'$ has side lengths AP, BP, CP, so we need to find the largest angle in triangle $\triangle BPP'$. As $\triangle APP'$ is equilateral we first have that

$$\angle PP'B = \angle APC - 60° = 124° - 60° = 64°.$$

Similarly

$$\angle P'PB = \angle APB - 60° = 113° - 60° = 53°.$$

Finally this means $\angle PBP' = 63°$ so we see the largest angle is $64°$.

Answer: 64

Problem 19 Solution

Let $y_1 = f(x)$ and $y_2 = f(-x)$, then $x = g(y_1)$ and $-x = g(y_2)$, and also $y_1 + y_2 = 4$. Then $g(y_1) = -g(4 - y_1)$ for arbitrary y_1. Since $(x-1) + (5-x) = 4$, the answer is $g(x-1) + g(5-x) = 0$.

Answer: 0

Problem 20 Solution

First note using patterns or Euler's Theorem,

$$47^4 \equiv 7^4 \equiv 1 \pmod{10}.$$

Further note that $47 \equiv 3 \pmod 4$, then

$$47^3 \equiv 3^3 \equiv 27 \equiv 3 \pmod 4,$$

so that

$$47^{47} \equiv 3^{47} \equiv 3^3 \equiv 3 \pmod 4.$$

Repeated use of this fact gives that

$$47^{47^{\cdot^{\cdot^{47}}}} \equiv 7^3 \equiv 343 \equiv 3 \pmod{10},$$

so our answer is 3. (Note in fact as long as there are at least 2 total 47s, the last digit is 3.)

Answer: 3

2.9 ZIML June 2017 Varsity

Below are the solutions from the Varsity ZIML Competition held in June 2017.

The problems from the contest are available on p.63.

Problem 1 Solution
Construct a circle with diameter AB. Then $\triangle ABE$ is obtuse if and only if point E is chosen inside this circle. As the total area of the square is 4 and the area of the circle inside the square (a semicircle) is $\pi/2$, then the probability is

$$\frac{\pi}{2} \div 4 = \frac{\pi}{8} \approx 0.3927,$$

so $K \approx 39.27$ and hence rounded to the nearest integer, is 39.

Answer: 39

Problem 2 Solution
Working mod 3, the Fibonacci numbers are

$$1, 1, 2, 0, 2, 2, 1, 0, 1, 1, \ldots$$

with the pattern $1, 1, 2, 0, 2, 2, 1, 0$ repeating every 8 terms. As $100 = 12 \times 8 + 4$, we see $12 \times 2 + 1 = 25$ of the first 100 Fibonacci numbers are a multiple of 3.

Answer: 25

Problem 3 Solution
Let x_i for $i = 1, 2, 3, \ldots, 13$ denote the 13 roots. Each of these roots satisfy $x_i^{13} - 13x_i + 13 = 0$, thus $x_i^{13} = 13x_i - 13$. By Vieta's formula,

$$\sum_{i=1}^{13} x_i = 0,$$

so

$$\sum_{i=1}^{13} x_i^{13} = \sum_{i=1}^{13} (13x_i - 13)$$

$$= 13 \left(\sum_{i=1}^{13} x_i \right) - 13 \cdot 13$$

$$= 0 - 169$$

$$= -169.$$

Answer: -169

Problem 4 Solution

Let $f(x) = x^4 + (m+n)x^3 + (m-n)x^2 + (m^2+2n-1)x+m+2$. Using long division (or the Method of Undetermined Coefficients) we can factor

$$f(x) = (x-1)(x^3 + (m+n+1)x^2 + (2m+1)x - (m+2)).$$

Let $g(x) = x^3 + (m+n+1)x^2 + (2m+1)x - (m+2)$. By the Factor Theorem, $f(1) = 0$ and $g(1) = 0$, and we have equations

$$\begin{cases} 1+m+n+m-n+m^2+2n-1+m+2 &= 0, \\ 1+m+n+1+2m+1-(m+2) &= 0. \end{cases}$$

Or after simplifying

$$\begin{cases} m^2+3m+2n+2 &= 0, \\ 2m+n+1 &= 0. \end{cases}$$

Substituting $n = -2m-1$ we get

$$m^2+3m-4m-2+2 = 0 \Rightarrow m(m-1) = 0 \Rightarrow m = 0,1.$$

If $m = 0$ then $n = -1$ and if $m = 1$ then $n = -3$. Hence our answer is $1 \times (-3) = -3$.

Answer: -3

Problem 5 Solution

First note that the fourth polygon must have an interior angle of $360° - 90° - 90° - 60° = 120°$ and hence must be a hexagon. Therefore the perimeter of the entire shape is the sum of the perimeter of each of the four polygons minus the 4 shared edges. Note each shared edge is counted twice. Hence the perimeter is $6 + 4 + 4 + 3 - 4 \cdot 2 = 9$.

Answer: 9

Problem 6 Solution

As $8 = 2^3$ we just need to count powers of 2. Start by making an educated guess. Trying 50!, it which contains $\left\lfloor \frac{50}{2} \right\rfloor + \left\lfloor \frac{50}{4} \right\rfloor + \left\lfloor \frac{50}{8} \right\rfloor + \left\lfloor \frac{50}{16} \right\rfloor + \left\lfloor \frac{50}{32} \right\rfloor = 25 + 12 + 6 + 3 + 1 = 47$ powers of 2. Therefore, 50! will end in $\left\lfloor \frac{47}{3} \right\rfloor = 15$ zeros when written in base 8 (as will 51! as 51 is odd). We need one more power of 2, so we try 52!, which contains $\left\lfloor \frac{52}{2} \right\rfloor + \left\lfloor \frac{52}{4} \right\rfloor + \left\lfloor \frac{52}{8} \right\rfloor + \left\lfloor \frac{52}{16} \right\rfloor + \left\lfloor \frac{52}{32} \right\rfloor = 26 + 13 + 6 + 3 + 1 = 49$ powers of 2, so the base 8 representation of 52! and 53! will end in $\left\lfloor \frac{49}{3} \right\rfloor = 16$ zeros. Similar calculations will show that 56! contains 53 powers of 2 and hence its base 8 representation ends in 17 zeros. Thus, $n = 52, 53, 54, 55$ are the four possibilities with $n!$ ending in 16 zeros.

Answer: 4

Problem 7 Solution

There are 1000000 numbers in total. Let A be the set of perfect squares, B perfect cubes, and C perfect fifth powers. Using inclusion and exclusion we have

$$
\begin{aligned}
n(A \cup B \cup C) &= n(A) + n(B) + n(C) - n(A \cap B) - n(A \cap C) \\
&\quad - n(B \cap C) + n(A \cap B \cap C) \\
&= 1000 + 100 + 15 - 10 - 3 - 2 + 1 \\
&= 1101.
\end{aligned}
$$

As this is the opposite of what we want, our final answer is $1000000 - 1101 = 998899$.

Answer: 998899

Problem 8 Solution

If d is a factor of n, then n/d must also be a factor of n. Therefore factors come in pairs whose product is n, except for the case where n is a perfect square (in which case $\sqrt{n}$ is an integer and paired with itself). Since the product of all factors is n^4, there are 4 pairs of distinct factors, thus n must have 8 factors. Since $8 = 4 \times 2 = 2 \times 2 \times 2$ numbers with 8 factors will be of the form p^7 or $p^3 \times q$ or $p \times q \times r$ where p, q, r are all prime numbers. Therefore the smallest will be either $2^7 = 128$ or $2^3 \times 3 = 24$ or $2 \times 3 \times 5 = 30$ so our answer is 24.

Answer: 24

Problem 9 Solution

First note
$$
\log_4(x^2) = \frac{\log_2(x^2)}{2} = \log_2(x).
$$

Hence a and b are solutions to $(\log_2(x))^2 - 3\log_2(x) - 1 = 0$. If $z = \log_2(x)$ we have that the sum of the roots of $z^2 - 3z - 1 = 0$ is 3. Hence $\log_2(a) + \log_2(b) = 3$. Thus, $\log_2(a \cdot b) = 3$ so $a \cdot b = 2^3 = 8$.

Answer: 8

Problem 10 Solution

The probability that the flip is heads and the sum is 7 is

$$\frac{1}{2} \times \frac{6}{36} = \frac{1}{12}.$$

The probability that the flip is tails and the sum is 7 is

$$\frac{1}{2} \times \frac{\binom{6}{2}}{216} = \frac{5}{144}.$$

Here we used the positive version of stars and bars ($a+b+c=7$ where $a,b,c \geq 1$) for the three dice rolls. Hence we have a final probability of

$$\frac{1}{12} + \frac{5}{144} = \frac{17}{144},$$

so $Q - P = 144 - 17 = 127$.

Answer: 127

Problem 11 Solution

Label the regions as in the figure below.

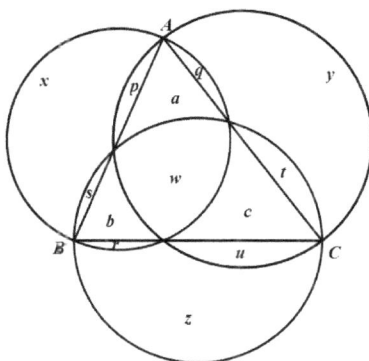

The area of $\triangle ABC$ is $a+b+c+w$. We are given that $x+y+z = 225$, and $w = 45$. Because $\overline{AB}, \overline{BC}, \overline{CA}$ are diameters, we have the following equations:

$$\begin{aligned} x+s+p &= a+q+w+b+r \\ y+q+t &= a+p+w+c+u \\ z+r+u &= b+s+w+c+t \end{aligned}$$

Adding these three equations,

$$\begin{aligned} &x+y+z+s+p+q+t+r+u \\ = \ &2(a+b+c)+3w+q+r+p+u+s+t. \end{aligned}$$

Therefore

$$x+y+z = 2(a+b+c+w)+w.$$

Thus

$$225 = 2(a+b+c+w)+45,$$

so

$$a+b+c+w = 90.$$

Answer: 90

Problem 12 Solution
First note that $3 \cdot 11 \cdot 31 = 1023 = 2^{10} - 1$. Any n between 0 and 1000 can be written as $10K + L$ for some integers K, L with $K \geq 0$ and $0 \leq L \leq 9$. Then

$$2^n - 1 \equiv (2^{10})^K \cdot 2^L - 1 \equiv 1 \cdot 2^L - 1 \pmod{1023}.$$

Hence we want $2^L - 1 \equiv 0 \pmod{1023}$ for $0 \leq L \leq 9$. This is clearly only true if $L = 0$. Therefore n of the form $10K$ all work, so we want the sum

$$10 + 20 + 30 + \cdots + 990 = \frac{(10 + 990) \cdot 99}{2} = 49500.$$

Answer: 49500

Problem 13 Solution

Find point K on side $\overline{BC}$ so that $\overline{DK} \parallel \overline{AB}$, and let $\overline{DK}$ intersects $\overline{EF}$ and $\overline{GH}$ at M and N, respectively.

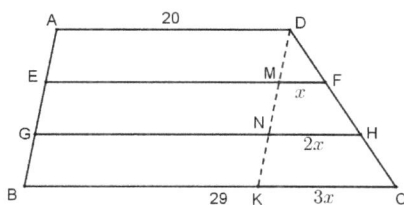

Then $\triangle DMF \sim \triangle DNH \sim \triangle DKC$ with side ratios $1:2:3$. Let $MF = x$, then $NH = 2x$ and $KC = 3x$. We also know that $KC = BC - BK = BC - AD = 29 - 20 = 9$, thus $x = 3$, so $EF = 23$.

Answer: 23

Problem 14 Solution

Note $1 = 1$ can be written in one way, $2 = 1 + 1 = 2$ in two ways, and $3 = 1 + 1 + 1 = 1 + 2 = 2 + 1 = 2$ in three ways. Look for a pattern in $1, 2, 3, 5, \ldots$ we guess the Fibonacci sequence (missing the first term). This gives:

$$1, 2, 3, 5, 8, 13, 21, 34, 55, 89, 144, 233, 377, 610$$

so our final answer is 610.

To prove this we claim that for any $n > 1$ the nth term in the Fibonacci sequence gives the number of ways $n - 1$ can be written as a sum of ones and twos. Then any sum of 1's and 2's either starts with a 1 or a 2. Therefore, removing the first term of a sequence summing to $n - 1$ gives either a sequence summing to $n - 2$ or $n - 3$. This shows that the sequence follows the same recursion as the Fibonacci sequence as claimed.

Answer: 610

Problem 15 Solution

The powers of 7 has a pattern in mod 96:

$$\begin{aligned}
7^1 &\equiv 7 \pmod{96}, \\
7^2 &\equiv 49 \pmod{96}, \\
7^3 &\equiv 55 \pmod{96}, \\
7^4 &\equiv 1 \pmod{96}, \\
7^5 &\equiv 7 \pmod{96},
\end{aligned}$$

$$\cdots$$

so there is a cycle of length 4. The exponent

$$7^{7^7} \equiv (-1)^{7^7} \equiv -1 \equiv 3 \pmod{4},$$

therefore the answer is $7^3 \equiv 55 \pmod{96}$.

Answer: 55

Problem 16 Solution

Since $(5, 12, 13)$ is a Pythagorean triple and $\dfrac{\pi}{2} < \theta < \pi$, we have
$\tan(\theta) = -\dfrac{5}{12}$. Hence

$$\tan(2x) = \frac{2\tan(x)}{1 - \tan^2(x)} = \frac{-120}{119},$$

so $P + Q = -120 + 119 = -1$.

Answer: -1

Problem 17 Solution

Using Angle Bisector Theorem on the angle bisectors BE and DE,

$$\frac{BC}{BA} = \frac{CE}{EA} = \frac{DC}{DA}.$$

By Law of Sines,

$$\frac{BC}{BA} = \frac{\sin\angle CAB}{\sin\angle C} \text{ and } \frac{DC}{DA} = \frac{\sin\angle CAD}{\sin\angle C} = \frac{\sin(\angle CAB/2)}{\sin\angle C}.$$

Thus $\sin \angle CAB = \sin \dfrac{\angle CAB}{2}$. Clearly $\angle CAB \neq \dfrac{\angle CAB}{2}$, so

$$\angle CAB = 180° - \frac{\angle CAB}{2},$$

then $\angle CAB = 120°$.

Answer: 120

Problem 18 Solution
Let $y = x - 2$ so $(y+2)^4 + (y-2)^4 = 626$. Note by symmetry the cubic and linear terms cancel when we expand. This gives

$$\begin{aligned} 2y^4 + 48y^2 + 32 &= 626, \\ y^4 - 24y^2 - 297 &= 0, \\ (y^2 + 33)(y^2 - 9) &= 0. \end{aligned}$$

Hence $y = \pm 3$ are the real roots, which gives $x = 5$ and $x = -1$ as real roots. The greatest root is $x = 5$, so $k = 5$.

Answer: 5

Problem 19 Solution
Let these points be $a_1, a_2, a_3, \ldots, a_{2016}$ in that order. Let M be the midpoint of line segment $\overline{a_1 a_{2016}}$. Then M is different from the midpoints of $\overline{a_2 a_{2016}}, \overline{a_3 a_{2016}}$, up to $\overline{a_{2015} a_{2016}}$ (a total of 2014 additional midpoints). Since all of these points are between M and a_{2016}, these are all different from the midpoints of $\overline{a_1 a_2}, \overline{a_1 a_3}$, up to $\overline{a_1 a_{2015}}$ (an additional 2014 midpoints). Hence there are at least $1 + 2014 + 2014 = 4029$ distinct midpoints.

To see we cannot guarantee more midpoints, consider $a_1, a_2, \ldots, a_{2016}$ equally spaced out (for example, $a_1 = (1,0)$, $a_2 = (2,0)$, etc.) from which it is easy to see that any other line segment shares a midpoint with one already listed.

Answer: 4029

Problem 20 Solution

As $x \neq 0$ we can divide by x to get $x + \dfrac{1}{x} = -1$. Further as $x^3 - 1 = (x-1)(x^2+x+1)$ (and hence a root of x^2+x+1 is a root of $x^3 - 1$) we know $x^3 = 1$.

Therefore $x^k + \dfrac{1}{x^k} = 2$ if k is a multiple of 3 and $x^k + \dfrac{1}{x^k} = -1$ otherwise.

Hence

$$x^{2016} + \frac{1}{x^{2016}} = 2,$$

$$x^{2017} + \frac{1}{x^{2017}} = -1,$$

$$x^{2018} + \frac{1}{x^{2018}} = -1,$$

and

$$\left(x^{2016} + \frac{1}{x^{2016}}\right)^3 + \left(x^{2017} + \frac{1}{x^{2017}}\right)^3 + \left(x^{2018} + \frac{1}{x^{2018}}\right)^3$$
$$= (2)^3 + (-1)^3 + (-1)^3$$
$$= 6.$$

Answer: 6

3. Appendix

3.1 Varsity Topics Covered

Algebra

- Students should be comfortable with ratios, proportions, and their applications to problems involving work and motion, but these problems are not a main focus at this level
- Radicals, Exponents, and Logarithms: Simplest Radical Form for Roots, Laws of Exponents, Laws of Logarithms including change of base
- Complex Numbers: Arithmetic Operations, Rectangular, Polar, and Exponential Forms, De Moivre's identity, Roots of Unity
- Factoring Tricks: Sums and differences of squares, cubes, etc., Binomial and multinomial theorem, Completing the Square/Rectangle, etc.
- Solving Equations: Linear Equations, Quadratic Equations, Systems of Equations, Substitutions to rewrite higher degree equations as quadratics, Radicals, Absolute Values
- Quadratics: Graphing and Vertex Form, Maxima and Minima, Quadratic Formula, Discriminant, Vieta's Theorem for sum and product of the roots

---

- Polynomials: Polynomial Long Division, Remainder and Factor Theorem, Rational Root Theorem, General Vieta's Theorem

Geometry

- As a general rule students should be comfortable using algebraic techniques (linear equations, quadratic equations, systems of equations, etc.) as tools for applying the geometric concepts listed below
- Angles in Parallel Lines (corresponding angles, alternating interior/exterior angles, same-side interior/exterior angles, etc.)
- Analytic Geometry: Equations of Lines, Parabolas, and Circles, Distance Formula, Midpoint Formula, Geometric Interpretation of Slope and Angles
- Triangles: Congruence and Similarity, Pythagorean theorem, Ratios of Sides for triangles with angles of 45, 45, 90 or 30, 60, 90
- Trigonometry: General understanding of sine, cosine, tangent, and their cofunctions, Law of Sines and Cosines, Trigonometric Identities for double angles, sums/differences, etc.
- Centers in Triangles: Definitions of altitudes, medians, angle bisectors, perpendicular bisectors, Definitions and basic properties of orthocenter, centroid, incenter, circumcenter, Angle Bisector Theorem, Ceva's and Menelaus's Theorems
- Interior and Exterior Angles of Polygons, including the sum of all these angles, each angle if the polygon is regular, etc.
- Areas and Perimeters of basic shapes such as triangles, rectangles, parallelograms, trapezoids, and circles, Heron's formula and formulas using inradius or circumradius for triangles

- Geometric Reasoning with Areas: Congruent shapes have the same area, Similar triangles have a ratio of areas that is the square of the ratio of their sides, Triangles with the same height have a ratio of their areas equal to the ratio of their bases, etc., Using multiple expressions of area to solve for unknowns
- Circles: Arc Length, Sector Area, Definitions for Tangent Lines and Tangent Circles, Inscribed Angles, Angles formed by intersecting chords, Power of a Point, Ptolemy's Theorem
- Lines and Plane in 3-D: Definitions of parallel, intersecting, and skew lines, Definitions of parallel and intersecting planes, Calculation of angles between lines and/or planes in 3-D space
- Solid Geometry: Surface Area and Volume for Spheres, Prisms, Pyramids, and Cones, Reasoning for more general solids, such as combining the solids listed above or pieces of solids when cut by a plane, etc.

Counting and Probability

- Fundamental Rules: Sum and Product Rules, Permutations and Combinations
- Counting Methods: Complementary counting, Stars and bars (also called sticks and stones, balls and urns, etc.), Grouping objects that must be together, Inserting objects that must be apart into spaces between objects, etc., Principle of Inclusion and Exclusion
- Identities: Symmetry, Pascal's Identity, Hockey Stick Identity, etc. for binomial coefficients, Binomial and Multinomial Theorem, Understanding of these identities using combinatorial proofs
- Sequences: Arithmetic and Geometric Sequences and Series, Finding and understanding patterns and recursive definitions for general sequences

- Probability and Sets: Definitions for event, sample space, complement, intersection, and union, Understanding the use of Venn Diagrams
- Probability: In finite sample spaces as a ratio of the number of outcomes, In geometric sample spaces as a ratio of lengths, areas, or volumes, Axioms of Probability, Independence, Conditional Probability, Law of Total Probability and Bayes's Theorem
- Probability Distributions: Definitions and Understandings of Probability Distributions and Expected Value

Number Theory

- Fundamental Definitions: Prime numbers, factors/divisors, multiples, least common multiple (LCM), greatest common factor/divisor (GCF or GCD), perfect squares/cubes/etc.
- Number Bases: Expressing and converting numbers in base 2, 3, 8, 16, etc, Understanding how to perform arithmetic in different bases
- Divisibility Rules for numbers such as 2, 3, 4, 5, 8, 9, 10, 11, and how to combine the rules for numbers such as 6, 22, etc.
- (Unique) Prime Factorization and how to use the prime factorization to find the number of factors, to test whether a number is a perfect square/cube/etc, to find the LCM or GCD.
- Factoring Tricks: Factors come in pairs, perfect squares have an odd number of factors, etc.
- Modular Arithmetic: Connection with remainders and applications such as "find the units digit", General rules for addition, subtraction, multiplication, and division, Extension of divisibility rules to calculating a number modulo 9, 11, etc., Fermat's Little Theorem, Euler's Totient Function and extension to Fermat's Little Theorem, Chinese Remainder Theorem

3.2 Glossary of Common Math Terms

Acute Angle An angle less than $90°$.

Altitude of a Triangle A line segment connecting a vertex of a triangle to the opposite side forming a right angle. Also called the height of a triangle.

Angle A figure formed by two rays sharing a common vertex. Often measured in degrees.

Angle Bisector A line dividing an angle into two equal halves.

Arc The curve of a circle connecting two points.

Area The amount of space a region takes up. Often denoted using square brackets: area of $\triangle ABC = [ABC]$.

Arithmetic Sequence A sequence where the difference between one term and the next is constant.

Average See Mean.

Base of a Triangle One side of a triangle, often used when the altitude is drawn from the opposite side to this base.

Binomial Coefficient The symbol $\binom{n}{k} = \dfrac{n!}{k!(n-k)!}$.

Centroid of a Triangle The intersection of the three medians in a triangle.

Chord A line segment connecting two points on the outside of a circle.

Circle A round shape consisting of points that all have the same distance (called the radius) from the center of the circle.

Circumcenter of a Triangle The intersection of the three perpendicular bisectors in a triangle. Also the center of the circle that circumscribes a triangle.

Circumference The perimeter of a circle.

Circumscribe To draw a shape outside another shape so that the boundaries touch.

Coefficient The number being multiplied by a variable or power of a variable. For example, the coefficient of x^3 in $5x^5 + 4x^3 + 2x$ is 4.

Complement In probability, the complement of a set is all elements outside the set.

Composite Number A number that is not prime.

Congruent Two shapes or figures that are exactly the same.

Cube A solid figure formed by 6 congruent squares that all meet at right angles.

Deck of Cards A standard deck of cards has 52 cards. There are 4 suits (clubs, diamonds, hearts, and spades) with each suit having cards of 13 ranks (A (ace), $2, 3, \ldots, 10$, J (jack), Q (queen), and K (king)).

Degree of a Polynomial The highest power of a variable in the polynomial. For example, the degree of $2x^3 - 5x^6 + 2$ is 6.

Denominator The bottom number in a fraction.

Diagonal A line segment connecting two vertices of a shape or solid that is not an edge of the shape or solid.

Diameter A chord passing through the center of a circle. The diameter has length that is twice the radius.

Die or Dice A standard die (plural is dice) has 6 sides. Each of the 6 sides has the same chance when the die is rolled.

Digit One of $0, 1, 2, \dots, 9$ used when writing a number.

Discriminant The expression $b^2 - 4ac$ for a quadratic equation $ax^2 + bx + c = 0$.

Distinguishable Objects Objects that are different.

Divisible A number is divisible by another number if there is no remainder when the first number is divided by the second. For example, 35 is divisible by 7.

Divisor A number that evenly divides another number. For example, 6 is a divisor of 48. Also called a factor.

Edge A line segment connecting two vertices on the outside of a shape or solid.

Equally Likely Having the same chance of occurring.

Equiangular Polygon A shape with all equal angles.

Equilateral Polygon A shape with all equal sides.

Equilateral Triangle A regular triangle, one with three equal sides and three equal angles.

Even Number A number divisible by 2.

Exponent The number another number is raised to for powers. For example, in a to the power of b (a^b), the exponent is b.

Exponential Form (of a complex number) A complex number written in the form $re^{i\theta}$ for real number r and angle θ.

Face The shape or polygon on the outside of a solid region.

Factor of a Number A number that evenly divides another number. For example, 6 is a factor of 48. Also called a divisor.

Factorial The symbol ! where $n! = n \times (n-1) \times (n-2) \cdots \times 1$.

Fraction An expression of a quotient. For example, $\dfrac{1}{2}$ or $\dfrac{9}{7}$.

Function A function is a rule that associates exactly one output with every input. Often described using an equation.

Geometric Sequence A sequence where the ratio between one term and the next is constant.

Greatest Common Divisor (GCD) The largest number that is a divisor/factor of two or more numbers.

Greatest Common Factor (GCF) See Greatest Common Divisor.

Incenter of a Triangle The intersection of the three angle bisectors in a triangle. Also the center of a circle that is inscribed inside a triangle.

Indistinguishable Objects Objects that are the same.

Inscribe To draw a shape inside another shape so that the boundaries touch.

Intersecting Lines or curves that cross each other.

Intersection of Two Sets The set of objects that are in both of the two sets. Denoted using ∩. For example, $\{2,3\} \cap \{3,4,5\} = \{3\}$.

Isosceles Triangle A triangle with two equal sides and two equal angles.

Least Common Multiple (LCM) The smallest number that is a multiple of two or more numbers.

Mean The sum of the numbers in a list divided by the how many numbers occur in the list. Also called the average.

Median The number in the middle of a list when the list is arranged in increasing order.

Median of a Triangle A line connecting a vertex in a triangle to the midpoint of the opposite side.

Midpoint The point in the middle of a line segment.

Mode The number or numbers occurring most often in a list of numbers.

Multiple A number that is an integer times another number. For example, 72 is a multiple of 8.

Numerator The top number in a fraction.

Obtuse Angle An angle between $90°$ and $180°$.

Odd Number A number not divisible by 2.

Orthocenter of a Triangle The intersection of the three altitudes in a triangle.

Parallel Lines Lines that do not intersect.

Perfect Cube A number that is another number cubed. For example, $64 = 4^3$ is a perfect cube.

Perfect Square A number that is another number squared. For example, $64 = 8^2$ is a perfect square.

Perimeter The length/distance around the outside of a shape.

Perpendicular Bisector A line perpendicular to and passing through the midpoint of a line segment.

Pi (π) A number used often in geometry. $\pi = 3.1415926\ldots \approx 3.14 \approx \dfrac{22}{7}$.

Polar Form (of a complex number) A complex number written in the form $r(\cos\theta + i\sin\theta)$ for real number r and angle θ.

Polygon A shape formed by connected line segments.

Polynomial A function that is made of adding multiples of powers of a variable. For example, $f(x) = x^4 + 3x^2 + 2x - 3$.

Prime Factorization The expression of a number as the product of all its prime factors. For example, 24 has prime factorization $2 \times 2 \times 2 \times 3 = 2^3 \times 3$.

Prime Number A number whose only factors are one and itself.

Proportional Ratios Ratios that have equal values when expressed in fraction form. For example, $2 : 3$ is proportional to $8 : 12$.

Quadratic A polynomial with degree 2. Often written in the form $ax^2 + bx + c$.

Quadrilateral A shape with four sides.

Quotient The integer quantity when dividing one number by another. For example, the quotient of $38 \div 5$ is 7 as $38 = 7 \times 5 + 3$.

Radius of a Circle The distance from the center of the circle to any point on the outside of the circle.

Randomly Chosen for a group of objects. Unless specified, the chance of choosing each object is the same as any other object.

Rank of a Card See Deck of Cards.

Ratio A relation depicting the relation between two quantities. For example $2 : 3$ or $\frac{2}{3}$ denotes that for every 3 of the second quantity there are 2 of the first quantity.

Rational Number A number that can be written as a fraction.

Reciprocal One divided by the number. For example, the reciprocal of 7 is $\frac{1}{7}$.

Rectangle A quadrilateral with four right angles (an equiangular quadrilateral).

Rectangular Form (of a complex number) A complex number written in the form $a + bi$ for real numbers a and b.

Regular Polygon A polygon with all equal sides and all equal angles (equilateral and equiangular).

Remainder The quantity left over when one integer is divided by another. For example, the remainder of $38 \div 5$ is 3 as $38 = 7 \times 5 + 3$.

Rhombus A quadrilateral with four equal sides (an equilateral quadrilateral).

Right Angle A $90°$ angle.

Right Triangle A triangle containing a right angle.

Root of a Function A value of x such that the function evaluates to zero. For example, $x = 2$ is a root of the function $f(x) = x^2 - 4$.

Sample Space In probability, the sample space is the set of all outcomes for an experiment.

Scalene Triangle A triangle with three unequal sides and three unequal angles.

Sector The region formed by an arc and the two radii connecting the ends of the arc to the center of the circle.

Sequence An ordered list of numbers.

Set An unordered collection or group of objects without repeated elements. Denoted using curly brackets. For example, $\{1,2,3,4\}$ is the set containing the integers $1,\ldots,4$.

Similar Shapes or solids that have the same angles and sides that share a common ratio.

Simplest Radical Form An expression containing a radical such that the number inside the radical is an integer that has no perfect squares.

Skew Line Lines in 3-D space that neither intersect nor are parallel.

Sphere A round solid consisting of points that all have the same distance (called the radius) from the center of the sphere.

Square A shape with four equal sides and four equal angles (a regular quadrilateral).

Subset A set of objects that is contained inside a larger set of objects. Denoted using $\subseteq$. For example $\{2,3\} \subseteq \{1,2,3,4\}$.

Suit of a Card See Deck of Cards.

Surface Area The total area of all the faces of a solid.

Tangent Line A line touching a shape or curve at exactly one point.

Trapezoid A quadrilateral with one pair of parallel sides.

Triangle A shape with three sides.

Union of Two Sets The set of objects that are in one or both of the two sets. Denoted using $\cup$. For example, $\{2,3\} \cup \{3,4,5\} = \{2,3,4,5\}$.

Venn Diagram A diagram with circles used to understand the relationship between overlapping sets.

Vertex The intersection of line segments, especially the intersection of sides or edges in a shape or solid.

Volume The amount of space a solid region takes up.

With Replacement When choosing objects with replacement, a chosen object is returned to the others allowing it to be chosen more than once.

3.3 ZIML Answers

ZIML October 2016 Varsity

Problem 1:	6	**Problem 11:**	11
Problem 2:	185	**Problem 12:**	128
Problem 3:	406	**Problem 13:**	60
Problem 4:	6965	**Problem 14:**	159
Problem 5:	12	**Problem 15:**	32
Problem 6:	5	**Problem 16:**	15
Problem 7:	9504	**Problem 17:**	11
Problem 8:	38	**Problem 18:**	2
Problem 9:	43	**Problem 19:**	700
Problem 10:	5	**Problem 20:**	15

ZIML November 2016 Varsity

Problem 1:	8	Problem 11:	3
Problem 2:	2	Problem 12:	7
Problem 3:	705	Problem 13:	46
Problem 4:	10	Problem 14:	81
Problem 5:	28	Problem 15:	23430
Problem 6:	5	Problem 16:	2017
Problem 7:	343	Problem 17:	-5
Problem 8:	900	Problem 18:	120
Problem 9:	28	Problem 19:	1.5
Problem 10:	2019	Problem 20:	1501

ZIML December 2016 Varsity

Problem 1:   1   Problem 11:   121

Problem 2:   768   Problem 12:   25

Problem 3:   7   Problem 13:   1008

Problem 4:   −∠   Problem 14:   14

Problem 5:   35964   Problem 15:   3

Problem 6:   13   Problem 16:   12

Problem 7:   1.5   Problem 17:   128

Problem 8:   5   Problem 18:   31·

Problem 9:   37.5   Problem 19:   3

Problem 10:   −2   Problem 20:   20202

ZIML January 2017 Varsity

Problem 1:	8	Problem 11:	45
Problem 2:	2	Problem 12:	17
Problem 3:	−23	Problem 13:	3600
Problem 4:	322560	Problem 14:	−8
Problem 5:	3	Problem 15:	0
Problem 6:	76	Problem 16:	−252
Problem 7:	407	Problem 17:	508032
Problem 8:	6	Problem 18:	15
Problem 9:	23	Problem 19:	−2
Problem 10:	144	Problem 20:	10

ZIML February 2017 Varsity

Problem 1: −16

Problem 2: 504

Problem 3: 4293

Problem 4: 1404

Problem 5: −8128

Problem 6: 36

Problem 7: −4

Problem 8: 64

Problem 9: 22

Problem 10: 119

Problem 11: 12

Problem 12: 3

Problem 13: 4

Problem 14: 29

Problem 15: 5

Problem 16: 31

Problem 17: 12

Problem 18: 13

Problem 19: 5880

Problem 20: −1

ZIML March 2017 Varsity

Problem 1:	100	Problem 11:	2
Problem 2:	0	Problem 12:	90
Problem 3:	52800	Problem 13:	6
Problem 4:	3	Problem 14:	1170
Problem 5:	9	Problem 15:	18
Problem 6:	5040	Problem 16:	54
Problem 7:	2	Problem 17:	9000
Problem 8:	332640	Problem 18:	0.25
Problem 9:	45	Problem 19:	366
Problem 10:	60	Problem 20:	4

ZIML April 2017 Varsity

Problem 1:	18	Problem 11:	128
Problem 2:	143	Problem 12:	5
Problem 3:	880	Problem 13:	8008
Problem 4:	10	Problem 14:	9
Problem 5:	24	Problem 15:	12
Problem 6:	43200	Problem 16:	5
Problem 7:	1	Problem 17:	2016
Problem 8:	-91	Problem 18:	3
Problem 9:	6	Problem 19:	50
Problem 10:	148	Problem 20:	3.14

ZIML May 2017 Varsity

Problem 1:   15   Problem 11:   8

Problem 2:   5   Problem 12:   5

Problem 3:   13440   Problem 13:   −8

Problem 4:   6   Problem 14:   23

Problem 5:   240   Problem 15:   31

Problem 6:   44286   Problem 16:   3

Problem 7:   43   Problem 17:   5

Problem 8:   4   Problem 18:   64

Problem 9:   65   Problem 19:   0

Problem 10:   9697   Problem 20:   3

ZIML June 2017 Varsity

Problem 1:	39	**Problem 11:**	90
Problem 2:	25	**Problem 12:**	49500
Problem 3:	-169	**Problem 13:**	23
Problem 4:	-3	**Problem 14:**	610
Problem 5:	9	**Problem 15:**	55
Problem 6:	4	**Problem 16:**	-1
Problem 7:	998899	**Problem 17:**	120
Problem 8:	24	**Problem 18:**	5
Problem 9:	8	**Problem 19:**	4029
Problem 10:	127	**Problem 20:**	6